Reviews of Environmental Contamination and Toxicology

VOLUME 223

For further volumes:
http://www.springer.com/series/398

Reviews of Environmental Contamination and Toxicology

Editor
David M. Whitacre

Editorial Board
Maria Fernanda, Cavieres, Valparaiso, Chile • Charles P. Gerba, Tucson, Arizona, USA
John Giesy, Saskatoon, Saskatchewan, Canada • O. Hutzinger, Bayreuth, Germany
James B. Knaak, Getzville, New York, USA
James T. Stevens, Winston-Salem, North Carolina, USA
Ronald S. Tjeerdema, Davis, California, USA • Pim de Voogt, Amsterdam, The Netherlands
George W. Ware, Tucson, Arizona, USA

Founding Editor
Francis A. Gunther

VOLUME 223

Coordinating Board of Editors

DR. DAVID M. WHITACRE, *Editor*
Reviews of Environmental Contamination and Toxicology

5115 Bunch Road
Summerfield, North Carolina 27358, USA
(336) 634-2131 (PHONE and FAX)
E-mail: dmwhitacre@triad.rr.com

DR. HERBERT N. NIGG, *Editor*
Bulletin of Environmental Contamination and Toxicology

University of Florida
700 Experiment Station Road
Lake Alfred, Florida 33850, USA
(863) 956-1151; FAX (941) 956-4631
E-mail: hnn@LAL.UFL.edu

DR. DANIEL R. DOERGE, *Editor*
Archives of Environmental Contamination and Toxicology

7719 12th Street
Paron, Arkansas 72122, USA
(501) 821-1147; FAX (501) 821-1146
E-mail: AECT_editor@earthlink.ne

ISSN 0179-5953
ISBN 978-1-4614-5576-9   ISBN 978-1-4614-5577-6 (eBook)
DOI 10.1007/978-1-4614-5577-6
Springer New York Heidelberg Dordrecht London

© Springer Science+Business Media New York 2013
This work is subject to copyright. All rights are reserved by the Publisher, whether the whole or part of the material is concerned, specifically the rights of translation, reprinting, reuse of illustrations, recitation, broadcasting, reproduction on microfilms or in any other physical way, and transmission or information storage and retrieval, electronic adaptation, computer software, or by similar or dissimilar methodology now known or hereafter developed. Exempted from this legal reservation are brief excerpts in connection with reviews or scholarly analysis or material supplied specifically for the purpose of being entered and executed on a computer system, for exclusive use by the purchaser of the work. Duplication of this publication or parts thereof is permitted only under the provisions of the Copyright Law of the Publisher's location, in its current version, and permission for use must always be obtained from Springer. Permissions for use may be obtained through RightsLink at the Copyright Clearance Center. Violations are liable to prosecution under the respective Copyright Law.
The use of general descriptive names, registered names, trademarks, service marks, etc. in this publication does not imply, even in the absence of a specific statement, that such names are exempt from the relevant protective laws and regulations and therefore free for general use.
While the advice and information in this book are believed to be true and accurate at the date of publication, neither the authors nor the editors nor the publisher can accept any legal responsibility for any errors or omissions that may be made. The publisher makes no warranty, express or implied, with respect to the material contained herein.

Printed on acid-free paper

Springer is part of Springer Science+Business Media (www.springer.com)

Foreword

International concern in scientific, industrial, and governmental communities over traces of xenobiotics in foods and in both abiotic and biotic environments has justified the present triumvirate of specialized publications in this field: comprehensive reviews, rapidly published research papers and progress reports, and archival documentations. These three international publications are integrated and scheduled to provide the coherency essential for nonduplicative and current progress in a field as dynamic and complex as environmental contamination and toxicology. This series is reserved exclusively for the diversified literature on "toxic" chemicals in our food, our feeds, our homes, recreational and working surroundings, our domestic animals, our wildlife, and ourselves. Tremendous efforts worldwide have been mobilized to evaluate the nature, presence, magnitude, fate, and toxicology of the chemicals loosed upon the Earth. Among the sequelae of this broad new emphasis is an undeniable need for an articulated set of authoritative publications, where one can find the latest important world literature produced by these emerging areas of science together with documentation of pertinent ancillary legislation.

Research directors and legislative or administrative advisers do not have the time to scan the escalating number of technical publications that may contain articles important to current responsibility. Rather, these individuals need the background provided by detailed reviews and the assurance that the latest information is made available to them, all with minimal literature searching. Similarly, the scientist assigned or attracted to a new problem is required to glean all literature pertinent to the task, to publish new developments or important new experimental details quickly, to inform others of findings that might alter their own efforts, and eventually to publish all his/her supporting data and conclusions for archival purposes.

In the fields of environmental contamination and toxicology, the sum of these concerns and responsibilities is decisively addressed by the uniform, encompassing, and timely publication format of the Springer triumvirate:

Reviews of Environmental Contamination and Toxicology [Vol. 1 through 97 (1962–1986) as Residue Reviews] for detailed review articles concerned with any aspects of chemical contaminants, including pesticides, in the total environment with toxicological considerations and consequences.

Bulletin of Environmental Contamination and Toxicology (Vol. 1 in 1966) for rapid publication of short reports of significant advances and discoveries in the fields of air, soil, water, and food contamination and pollution as well as methodology and other disciplines concerned with the introduction, presence, and effects of toxicants in the total environment.

Archives of Environmental Contamination and Toxicology (Vol. 1 in 1973) for important complete articles emphasizing and describing original experimental or theoretical research work pertaining to the scientific aspects of chemical contaminants in the environment.

Manuscripts for Reviews and the Archives are in identical formats and are peer reviewed by scientists in the field for adequacy and value; manuscripts for the *Bulletin* are also reviewed, but are published by photo-offset from camera-ready copy to provide the latest results with minimum delay. The individual editors of these three publications comprise the joint Coordinating Board of Editors with referral within the board of manuscripts submitted to one publication but deemed by major emphasis or length more suitable for one of the others.

Coordinating Board of Editors

Preface

The role of *Reviews* is to publish detailed scientific review articles on all aspects of environmental contamination and associated toxicological consequences. Such articles facilitate the often complex task of accessing and interpreting cogent scientific data within the confines of one or more closely related research fields.

In the nearly 50 years since *Reviews of Environmental Contamination and Toxicology* (formerly *Residue Reviews*) was first published, the number, scope, and complexity of environmental pollution incidents have grown unabated. During this entire period, the emphasis has been on publishing articles that address the presence and toxicity of environmental contaminants. New research is published each year on a myriad of environmental pollution issues facing people worldwide. This fact, and the routine discovery and reporting of new environmental contamination cases, creates an increasingly important function for *Reviews*.

The staggering volume of scientific literature demands remedy by which data can be synthesized and made available to readers in an abridged form. *Reviews* addresses this need and provides detailed reviews worldwide to key scientists and science or policy administrators, whether employed by government, universities, or the private sector.

There is a panoply of environmental issues and concerns on which many scientists have focused their research in past years. The scope of this list is quite broad, encompassing environmental events globally that affect marine and terrestrial ecosystems; biotic and abiotic environments; impacts on plants, humans, and wildlife; and pollutants, both chemical and radioactive; as well as the ravages of environmental disease in virtually all environmental media (soil, water, air). New or enhanced safety and environmental concerns have emerged in the last decade to be added to incidents covered by the media, studied by scientists, and addressed by governmental and private institutions. Among these are events so striking that they are creating a paradigm shift. Two in particular are at the center of ever increasing media as well as scientific attention: bioterrorism and global warming. Unfortunately, these very worrisome issues are now superimposed on the already extensive list of ongoing environmental challenges.

The ultimate role of publishing scientific research is to enhance understanding of the environment in ways that allow the public to be better informed. The term "informed public" as used by Thomas Jefferson in the age of enlightenmentconveyed the thought of soundness and good judgment. In the modern sense, being "well informed" has the narrower meaning of having access to sufficient information. Because the public still gets most of its information on science and technologyfrom TV news and reports, the role for scientists as interpreters and brokers of scientificinformation to the public will grow rather than diminish. Environmentalismis the newest global political force, resulting in the emergence of multinational consortiato control pollution and the evolution of the environmental ethic.Will the new-politics of the twenty-first century involve a consortium of technologists and environmentalists,or a progressive confrontation? These matters are of genuine concernto governmental agencies and legislative bodies around the world.

For those who make the decisions about how our planet is managed, there is anongoing need for continual surveillance and intelligent controls to avoid endangeringthe environment, public health, and wildlife. Ensuring safety-in-use of the manychemicals involved in our highly industrialized culture is a dynamic challenge, forthe old, established materials are continually being displaced by newly developedmolecules more acceptable to federal and state regulatory agencies, public healthofficials, and environmentalists.

Reviews publishes synoptic articles designed to treat the presence, fate, and, ifpossible, the safety of xenobiotics in any segment of the environment. These reviewscan be either general or specific, but properly lie in the domains of analytical chemistryand its methodology, biochemistry, human and animal medicine, legislation,pharmacology, physiology, toxicology, and regulation. Certain affairs in food technologyconcerned specifically with pesticide and other food-additive problems mayalso be appropriate.

Because manuscripts are published in the order in which they are received infinal form, it may seem that some important aspects have been neglected at times.However, these apparent omissions are recognized, and pertinent manuscripts arelikely in preparation or planned. The field is so very large and the interests in itare so varied that the editor and the editorial board earnestly solicit authors andsuggestions of underrepresented topics to make this international book series yetmore useful and worthwhile.

Justification for the preparation of any review for this book series is that it dealswith some aspect of the many real problems arising from the presence of foreignchemicals in our surroundings. Thus, manuscripts may encompass case studies fromany country. Food additives, including pesticides, or their metabolites that may persistinto human food and animal feeds are within this scope. Additionally, chemicalcontamination in any manner of air, water, soil, or plant or animal life is within theseobjectives and their purview.

Manuscripts are often contributed by invitation. However, nominations for newtopics or topics in areas that are rapidly advancing are welcome. Preliminary communicationwith the editor is recommended before volunteered review manuscriptsare submitted.

Summerfield, NC, USA   David M. Whitacre

Contents

**Air Contaminant Statistical Distributions
with Application to PM10 in Santiago, Chile** 1
Carolina Marchant, Víctor Leiva, M. Fernanda Cavieres,
and Antonio Sanhueza

**Advances in the Application of Plant Growth-Promoting
Rhizobacteria in Phytoremediation of Heavy Metals** 33
Hamid Iqbal Tak, Faheem Ahmad,
and Olubukola Oluranti Babalola

**Toxicity Reference Values and Tissue Residue
Criteria for Protecting Avian Wildlife Exposed
to Methylmercury in China** 53
Ruiqing Zhang, Fengchang Wu, Huixian Li, Guanghui Guo,
Chenglian Feng, John P. Giesy, and Hong Chang

**The Biological Effects and Possible Modes
of Action of Nanosilver** 81
Carolin Völker, Matthias Oetken, and Jörg Oehlmann

**Diazinon—Chemistry and Environmental Fate:
A California Perspective** 107
Vaneet Aggarwal, Xin Deng, Atac Tuli, and Kean S. Goh

Erratum E1

Index 141

Air Contaminant Statistical Distributions with Application to PM10 in Santiago, Chile

Carolina Marchant, Víctor Leiva, M. Fernanda Cavieres, and Antonio Sanhueza

Contents

1	Introduction	2
2	Air Contaminants and Their Health Effects	5
	2.1 Gaseous Compounds	5
	2.2 Heavy Metals	7
	2.3 Persistent Organic Pollutants	7
	2.4 Suspended Solid Particles	7
3	Air Contaminant Distributions	8
	3.1 Larsen (1971)'s Pioneering Work	8
	3.2 After Larsen's Work	8
	3.3 Ott (1990)'s Physical Justification	12
	3.4 Birnbaum–Saunders Air Contaminant Distribution (2008)	12
4	Air Contamination in Santiago, Chile	12
	4.1 Location	12
	4.2 Wind and Thermal Inversion Effects	14
	4.3 Monitoring Stations and Data Measurement	14
	4.4 Human Health Effects and International Guidelines	15
	4.5 A Study in Santiago, Chile	16

C. Marchant • V. Leiva (✉)
Departamento de Estadística, Universidad de Valparaíso,
Gran Bretaña 1111, Playa Ancha, Valparaíso, Chile
e-mail: victor.leiva@uv.cl

M.F. Cavieres
Facultad de Farmacia, Universidad de Valparaíso, Valparaíso, Chile

A. Sanhueza
Departamento de Matemática y Estadística, Universidad de La Frontera, Temuco, Chile

5	Application to PM10 Concentrations in Santiago, Chile	17
	5.1 Problem Under Analysis and Data	17
	5.2 Autocorrelation Analysis	18
	5.3 Exploratory Data Analysis	18
	5.4 Selection and Validation of the Model	20
	5.5 Computing Exceedance Probabilities and Percentiles	23
	5.6 Connecting Statistical Information with Adverse Effects on Human Health	24
6	Conclusions	25
7	Summary	26
References		27

1 Introduction

Breathable air is a gas mixture made up of 78% nitrogen, 21% oxygen, and 1% carbon dioxide and other gases such as argon, radon, and xenon (Pani 2007). Atmospheric contamination is the presence in the air of substances that change their chemical and physical characteristics. Air pollution derives primarily from fossil fuel combustion products that are emitted into the air. In some areas, the effects of air pollution are exacerbated when climatological and geographical factors restrict its dissipation. Over the past decades, the air quality of many urban centers has seriously deteriorated. As a result, millions of people are exposed to pollution levels above the World Health Organization (WHO) limits, as indicated by the United Nations Environment Programme. Air pollution is currently a concern in the American region, wherein several capital cities have levels that exceed national and international guideline limits. Such is the case for Santiago, the capital city of Chile, which is among the cities with higher air pollution levels in the world (Ostro 2003). The location of Santiago and the weather it experiences, when combined with high anthropological emissions, create critical air pollution conditions. The interaction of air pollution and heat can impair the health and well-being of people, particularly the elderly and children (Kinney 2008).

Exposure to air contaminants causes diverse effects on human health. Such effects vary according to the nature of the contaminant, the concentration and duration of exposure to the contaminant, and synergistic activity among air pollutants. Susceptibility of the exposed population is also a factor that influences the severity of adverse effects, especially in children, aged people, and chronic respiratory patients. Because of the multicomponent nature of contaminated air, it is difficult to establish a specific causal agent among the pollutant mixture that produces negative health effects. Numerous epidemiological studies have shown that air pollution causes chronic health problems and spreads infectious disease (e.g., diarrhea and intestinal worm infections). In addition, air pollution enhances or produces cardiovascular and pulmonary morbidity and increases mortality rates after exposure, particularly among children of lower income families, as indicated by the Organization for Economic Cooperation and Development (OECD), www.oecd.org; the Development Assistance Committee (DAC), www.oecd.org/dac; and by Satterthwaite (1997), Listorti (1999), and OECD-DAC (2000). Other epidemiological associations that have been established between disease and air pollution include cancer and reproductive and immunological alterations (WHO 2005; Kampa and Castanas 2008).

Annual deaths in the American region resulting from air pollution may be as high as 93,000 from cardiopulmonary diseases and 13,000 from lung cancer. There may also be a total of 58,000 life-years lost from acute respiratory infections in children who are under 4 years of age, plus a total of 560,000 adult disability-adjusted life years (Cohen et al. 2004). The serious effects of air pollution on public health can be seen not only in terms of disease and death, but also in terms of decreased productivity, missed education, and other reduced opportunities for human development (Cifuentes et al. 2001).

Statistical models can be used to describe air quality and these can be beneficial in the study of the relative impact of atmospheric contaminants on human health and on the urban environment. In epidemiological (monitoring) studies, average air pollutant concentrations have been used as indicators of representative air contamination levels that produce certain associated adverse human effects, e.g., chronic bronchitis (McConnell et al. 2003; Nuvolone et al. 2011). Periodic episodes of extreme concentrations of air pollution sometimes occur with certain atmospheric contaminants. Such episodes and their associated high concentrations vary with geographical and meteorological fluctuations and depend on changes in both source and type of emissions. Because of this variation, the air pollutant concentrations are treated as non-negative random variables that can be modeled by a statistical distribution. Generally, this distribution is asymmetrical and presents positive skewness (asymmetry). Therefore, the most popular statistical distribution (viz., the well-known normal or Gaussian distribution) cannot be applied because it is symmetrical. In some cases, environmental researchers transform their data to eliminate the skewness, and to be able to use the normal distribution. However, it has been shown that analysis performed under an inappropriate data transformation reduces the power of the study (Leiva et al. 2009 and references therein). In any case, even when an appropriate transformation is used, there still remains a problem of data interpretation. An alternative approach to avoid the transformation step is to model the data directly through an adequate statistical distribution.

Air contaminants are generally classified into four categories: (1) gaseous compounds such as carbon monoxide (CO), nitrogen dioxide (NO_2), sulfur dioxide (SO_2), tropospheric ozone (O_3), and volatile organic compounds (VOC); (2) heavy metals such as cadmium, lead, and mercury; (3) persistent organic pollutants such as dioxins, polychlorinated biphenyls (PCBs), and polycyclic aromatic hydrocarbons (PAHs); and (4) suspended particles or particulate matter (PM) such as PM2.5, and PM10. In particular, various effects are produced from exposure to PM, but the nature of those induced effects vary according to the PM composition (Brook et al. 2004; WHO 2005). Indeed, there is evidence for an increase in the risk of cardiovascular diseases and mortality from exposure to PM2.5, which occurs even after short time periods, such as hours or weeks (Brook et al. 2010). Maggiora and Lopez-Silva (2006) analyzed the levels of air pollution occurring in urban centers in Latin America and the Caribbean. They found that in 45 out of 53 urban centers, PM exceeded allowable levels. PM is a frequently monitored pollutant that is associated with the greatest risks to human health. (A well-known database called NMMAPS—www.ihapss.jhsph.

edu/data/NMMAPS/descriptives/frame.htm—shows that PM2.5 has the highest rate of missing data among air pollutants. PM10 is in the same situation, but the amount of missing data is not as high as for PM2.5.) Moreover, air quality varies within a given city. People who live near roads with high traffic density usually have the highest exposure level (Brauer et al. 2003; van Roosbroeck et al. 2006; Soliman et al. 2006). Such is the case for people living in Santiago, Chile. Due to variations in composition, PM10 concentrations are often considered as random variables that are associated with a statistical distribution. Thus, there are several reasons for considering PM10 concentrations as random variables and studying such variables.

Official environmental guidelines utilize an air contaminant concentration distribution to compute exceedance probabilities and percentiles that allow for the determination of administrative targets or establishing regulatory environmental alerts. Such alerts can protect human health at certain sites and times, because they address episodes of extreme air pollution that require corrective measures. Indeed, even standard measures applied to address the usual levels of air pollution provide considerable benefits. For example, the control measures currently applied to air pollution in Santiago, Chile, mainly limit the circulation of private vehicles and monitor the emission quality. Such efforts may well prevent deaths among the most vulnerable exposed groups (Cifuentes et al. 2001). Nevertheless, although several statistical distributions may appropriately fit pollutant concentrations at their central range, it is necessary to precisely establish what distribution of air pollutant concentrations fits better in the range of higher concentrations. For example, the air quality regulatory guideline of the Chilean government, established by the Ministry of the Environment (www.mma.gob.cl), formerly called National Commission for the Environment (CONAMA—Comisión Nacional del Medio Ambiente—in Spanish), indicates that any pollution value for PM10 that exceeds 150 µg/normalized cubic meters ($\mu g/m^3 N$) over a 24-h period is not allowed. Specifically, this guideline is surpassed when the 98th percentile of the PM10 24-h concentration collected at any monitoring station is greater than 150 $\mu g/m^3 N$. The Chilean guideline also warns that chronic health effects may be produced by PM10 concentrations during prolonged exposures; such chronic effects are manifested by an increment of the incidence and severity of diseases such as asthma, bronchitis, and emphysema (CONAMA 1998).

Several distributions have been used to model air pollutant concentrations. A number of researchers have utilized the log-normal (LN) distribution, mainly because of its theoretical properties and its relationship with the normal distribution (Larsen 1971; Ott and Mage 1976; Bencala and Seinfeld 1976; Ott 1990, 1995; Nevers 2000; Rumburg et al. 2001; Lu and Fang 2002; Leiva et al. 2008). Other distributions that have been applied to model air pollutant concentrations are the beta, exponential, extreme values (EV), gamma, inverse beta, inverse Gaussian (IG), Johnson SB (JSB), log-logistic (LL), Pearson, and Weibull models (Barry 1971; Scriven 1971; Singpurwalla 1972; Gifford 1974; Lynn 1974; Curran and Frank 1975; Pollack 1975; Tsukatani and Shigemitsu 1980; Berger et al. 1982; Holland and Terence 1982; Simpson et al. 1984; Taylor et al. 1986; Morel et al. 1999; Kan and Chen 2004; Sedek et al. 2006; Gokhale and Khare 2007; Nadarajah 2008; Deepa and Shiva 2010; Vilca et al. 2010).

One distribution that is derived from physical considerations of material failure due to fatigue is the Birnbaum-Saunders (BS) distribution (Birnbaum and Saunders 1969; Johnson et al. 1995, pp. 651–613; Leiva et al. 2007). This distribution has also been successfully applied to describe air pollutant concentrations (Leiva et al. 2008, 2010; Vilca et al. 2010, 2011; Ferreira et al. 2012). The use of the BS distribution has been justified by means of theoretical arguments. This distribution has suitable properties that are similar to those of the LN distribution.

All of the above-mentioned distributions, including the BS one, have parameter estimates that are sensitive to atypical (extreme) data, a frequent situation produced when one analyzes air pollutant concentrations. Díaz-García and Leiva (2005) developed a modification of the BS distribution known as the generalized BS (GBS) distribution which may be more appropriate for describing the phenomenon of air contamination. The parameter estimates obtained from this distribution are known to be robust to atypical data (Vilca and Leiva, 2006; Sanhueza et al. 2008; Balakrishnan et al. 2009). Section 3 provides more details about how these distributions may be specifically applied.

The aims of this chapter are to:

1. Provide a notion of the seriousness of air contaminants on human health.
2. Review the statistical distributions that have been used to model air contaminant data.
3. Describe the air contamination problems that occur in Santiago, Chile.
4. Propose a statistical methodology that can be applied for modeling air quality. We exemplify this proposal using previously unpublished air contamination data from the city of Santiago, Chile.

After addressing the general aspects of air contamination and its associated health effects in Sects. 1 and 2, we review the statistical distributions that are used for modeling air quality in Sect. 3. In Sect. 4, we describe the environmental problems faced by Santiago, Chile, and address aspects of Chilean and international regulation of air quality. In Sect. 5, we show how the BS distribution methodology can be applied to air contamination in this Chilean city. Finally, in Sect. 6, we sketch some conclusions that could facilitate reader's comprehension of the results presented in this study.

2 Air Contaminants and Their Health Effects

Below, we discuss general aspects of atmospheric contaminations and their effects on human health.

2.1 Gaseous Compounds

The primary gaseous air pollutants are CO, the dioxides of nitrogen or sulfur, ozone, and several VOCs.

CO is produced by combustion of carbonaceous materials. Its contribution to air pollution derives mainly from increased motor vehicle use and from road congestion, such as indicated by the United Nations Economic Commission for Latin America and the Caribbean (CEPAL—Comisión Económica para América Latina y el Caribe—in Spanish, 2000). Although unleaded gasoline has been available for decades in many countries of the American region, vehicle emissions still produce 80–90% of the lead present in the environment (World Bank 2001). Minor sources of lead emitted into the air include cigarette smoke, indoor heating systems, and kitchen appliances. In humans, inhaled CO is rapidly absorbed. Once in the blood, CO binds to hemoglobin and prevents oxygen binding. Thus, CO interferes with oxygen transport to tissues, affecting brain and heart functions (Katsouyanni 2003).

NO_2 can be distinguished by its yellow-brown color. This contaminant is water soluble and is produced by high temperature fuel combustion and volcanic eruptions. NO_2 is an important ozone precursor, since it reacts with organic volatiles under solar light conditions. NO_2 also contributes to the production of acid rain, thus increasing PM levels. Researchers have reported epidemiological associations between NO_2 exposure and pulmonary dysfunction, including asthma. However, there is no clear-cut threshold that delineates what constitutes harmful levels, especially after short-term exposures (Hesterberg et al. 2009; Latza et al. 2009).

SO_2 is a combustion product of sulfur-containing fossil fuels such as coal, gasoline, and other petroleum products. This contaminant is also emitted from metal smelters and other industrial processes and from volcano emissions. SO_2 is a potent constrictor of airways, even after only a few minutes of exposure. Its effect is greater in winter time, because bronchial reactivity increases when breathing cold and dry air. During the atmospheric oxidation process, SO_2 forms sulfates that can be transported as PM10, which, in the presence of humidity, creates acids that increase PM levels. Hence, SO_2 is not only an air contaminant itself, but is also a precursor of PM. Exposure to SO_2 can produce both acute and chronic health effects, including heart and lung morbidity and mortality (Hedley et al. 2002; Kan et al. 2010).

O_3 is a photochemically induced secondary pollutant generated by the interaction between nitrogen oxides and hydrocarbons under solar light conditions. Photochemical processes associated with the formation of ozone are highly complex and depend on the proportion in air of nitrogen oxides and different types of hydrocarbons. O_3 remains in the atmosphere for a long time, increasing in the summer months from high amounts of solar radiation. O_3 exposure in humans irritates the respiratory system, producing asphyxia, coughing, lung problems, and pneumonia. O_3 can also increase the severity of chronic respiratory diseases including asthma, bronchitis, and emphysema (WHO 2005).

VOCs are volatile at ambient temperatures and a very heterogeneous group of carbon compounds. The VOCs are precursors of O_3 as a result of their interaction with nitrogen oxides. Depending on their nature, VOCs may produce several different health effects including irritation of airways and skin, and cancer (Cicolella 2008).

2.2 Heavy Metals

Metals are ubiquitous in the environment and are emitted into the air from mining and other anthropogenic activities, as well as from the combustion of fossil fuels. Heavy metals are found in air pollution primarily adsorbed onto PM, and include elements such as arsenic, cadmium, lead, and mercury. In humans, heavy metals induce a variety of toxic effects including cancer and neurological impairment (Järup 2003), the latter effect probably deriving from its capacity to mediate oxidative stress and apoptotic reactions (Franco et al. 2009).

2.3 Persistent Organic Pollutants

These contaminants are made up of a very heterogeneous group of carbon compounds, such as dioxins, PAHs, and PCBs. They are generally very lipophilic and are not transformed by enzymatic reactions in organisms or by natural weathering. These compounds persist in organisms and in the environment for long periods of time. Dioxins, the PAHs, and the PCBs have been extensively studied for their environmental ubiquity in substrates, including food. Furthermore, these compounds are adsorbed onto PM, which facilitates inhalation exposure to them. Experimental data have shown that members of these persistent molecules (PAHs) produce genotoxicity, interact with cytoplasmic and nuclear receptors (dioxins, PAHs and PCBs), and may mediate a variety of effects that include cancer and reproductive defects (Schecter et al. 2006; Liu et al. 2008; White and Birnbaum 2009).

2.4 Suspended Solid Particles

The nature of PM material varies greatly, and includes heavy metals, silicates, sulfates, and a variety of hydrocarbons. PM is classified according to its diameter, because particle size determines sites of deposition within the respiratory tract. Coarser particles (those with a diameter over 10 μm) do not penetrate into airways. Instead, these particles are deposited in the upper respiratory tract and are cleared by cilia action. Inhalable particles are those measuring less than 10 μm; accordingly, they are called PM10 and can be further classified into particulates that are larger and others smaller than 2.5 μm (PM2.5). As size decreases, there is a higher possibility for PM to penetrate deeper into smaller alveoli and airways (WHO 2005).

Size also influences particle behavior in the atmosphere. For example, smaller particles tend to stay suspended in air for longer periods of time allowing them to travel hundreds of kilometers. In contrast, larger sized particles tend to be deposited closer to their source of origin. Although health effects of PM vary according to particle composition, which depends on location and time, one can say that acute and chronic exposures have been epidemiologically linked to a series of the following health-related problems: asthma, atherosclerosis, cancer, cardiovascular morbidity, diabetes, hospital admissions, and mortality (WHO 2005). Recently, the American Heart Association has issued an update to its 2004 scientific statement on air pollution and cardiovascular diseases (Brook et al. 2004), and provided additional evidence to show that an increased risk of cardiovascular diseases and mortality results from exposure to PM2.5. Moreover, such effects occur after short-term exposures of even hours or weeks (Brook et al. 2010).

3 Air Contaminant Distributions

Below, we review the diverse statistical distributions that have been used for modeling air quality.

3.1 Larsen (1971)'s Pioneering Work

Larsen (1971) published the results of a pioneer study in which he used statistical distributions different from the Gaussian one for modeling atmospheric pollutants. Larsen's data were collected between the years 1961 and 1968 for some contaminant agents (CO, hydrocarbons, NO_2, NO_x, O_3, SO_2) in eight US cities (Chicago, Cincinnati, Denver, Los Angeles, Philadelphia, St. Louis, San Francisco, and Washington). He concluded that the two-parameter LN distribution always showed the best fit in a manner that was independent of the type of contaminant or conditions of the sampled site.

3.2 After Larsen's Work

Singpurwalla (1972) applied the EV distribution to Larsen's data, proposing a new methodology based on the maximum contaminant concentrations.

Following the work of Larsen and Singpurwalla, several other authors used statistical distributions different from the LN and EV distributions to describe air contaminant concentrations (Barry 1971; Scriven 1971; Gifford 1974; Lynn 1974; Curran and Frank 1975; Pollack 1975). These distributions included the beta, exponential, gamma, three-parameter LN, Pearson, and Weibull distributions. In most of these studies, the two-parameter gamma and LN distributions showed better fit than the three- and four-parameter distributions. More details on these studies are given in Table 1. Tsukatani and Shigemitsu (1980) showed that the LN distribution is not always appropriate for modeling air pollutant data. They mentioned it is better to use models with heavier tails.

Table 1 Details of key air contaminant studies performed around the world

Type of contaminant	Site and year of collected data	Comments on each statistical study	References
CO,[a] hydrocarbons, NO_2,[b] NO_x,[c] O_3,[d] and SO_2[e]	Chicago, Cincinnati, Denver, Los Angeles, Philadelphia, St. Louis, San Francisco, Washington 1961–1968	Pioneer study using statistical analysis for modeling atmospheric pollutants. The two-parameter LN[f] distribution always showed the best fit, independently of the type of contaminant or site for collecting	Larsen (1971)
		The author applied EV[g] analysis to Larsen's data and proposed a new statistical tool, which does not demand collecting and analyzing a huge amount of data, but instead requires only the distribution of maximum contaminant concentrations	Singpurwalla (1972)
		The authors hypothesized that the two- and three-parameter LN, gamma and Weibull distributions could be used to model the CO, data previously analyzed by Larsen (1971). They indicated that CO concentrations are the result of complex phenomena which cannot be exactly determined and that the LN distribution could be used for a practical approach. They further went on to show that the three-parameter LN model gave a better fit to Larsen's data	Bencala and Seinfeld (1976)
CO, hydrocarbons NO_2, O_3, PM,[h] and SO_2	Germany (Frankfurt, Upland), Denmark (Copenhagen), US (Cincinnati, Los Angeles, St. Louis, Washington) and France 1961–1968	The authors compared the two-parameter LN distribution to the censored three-parameter LN distribution and concluded that the latter was the best option for describing atmospheric phenomena	Ott and Mage (1976)
SO_2	Four monitoring stations in the industrial coast region of Sakai, Japan, and eight stations located around the Misaki power central plant 1971–1972	The authors used the Pearson distribution system, which includes the beta, gamma, and inverted beta models, for describing SO_2 concentration data. They concluded that this distribution system provided a good fit for a wide variety of data and it is more flexible than the LN distribution	Tsukatani and Shigemitsu (1980)
SO_2	Twelve monitoring stations in Ghent, Belgium 1977–1979	The authors concluded that the two-parameter exponential distribution had an excellent fit when used to describe high concentrations of contaminants, while the two-parameter gamma distribution was the best model used to analyze the whole data set (when compared to the LN distribution)	Berger et al. (1982)

(continued)

Table 1 (continued)

Type of contaminant	Site and year of collected data	Comments on each statistical study	References
O_3	Chequamegon National Forest in the state of Wisconsin, US July and August, 1979	The four-parameter beta, three-parameter gamma, JSB,[i] three-parameter LN, normal, and three-parameter Weibull distributions were applied to the data set. The authors stated that there are many factors that have to be considered when selecting a statistical distribution for air quality. They concluded that the three-parameter LN model had a better fit than the other distributions. They additionally suggested that the truncated LN model could also be appropriate	Holland and Terence (1982)
SO_2	A central power plant in Upper Hunter Valley in New South Wales, Australia 1980–1981	The authors proposed using the exponential, two- and three-parameter LN, and Weibull distributions to model maximum concentrations of pollutants	Simpson et al. (1984)
CO, NO_2, NO_x, O_3, PM, and SO_2	Melbourne, Australia 1975–1984	The authors concluded that the LN distribution was appropriate for NO_2, PM, and SO_2 data, while the gamma distribution had a better fit for CO, NO_x, and O_3. The Weibull distribution was adequate for CO and O_3 data	Taylor et al. (1986)
NO_2, PM2.5,[j] PM10,[k] and SO_2	Santiago, Chile	The authors developed a new statistical distribution that related the level of emissions and the level of contamination. Thus, decreasing levels of emission could lead to better compliance of air quality guidelines; see Table 4. A fundamental assumption for this new distribution is that concentrations of air contaminant agents are the result of combinatorial emissions and random events. Other distributions used in this study were the beta, gamma, LN, Pearson type V, and Weibull models. The conclusion is that the Pearson type V model had a better fit to NO_2, PM2.5, PM10, and SO_2 data	Morel et al. (1999)
PM2.5 and PM8[l]	Spokane, Washington 1995–1997	The authors applied the EV, GEV,[m] gamma, two- and three-parameter LN, and Weibull distributions. The three-parameter LN distribution modeled the PM2.5 data appropriately, whereas the GEV distribution resulted to be more appropriate to the PM8 data	Rumburg et al. (2001)

(continued)

Table 1 (continued)

Type of contaminant	Site and year of collected data	Comments on each statistical study	References
PM2.5 and PM10	Sha-Lu, Taiwan July and December, 2000	The authors considered that the contaminant distribution varies with the site's meteorological factors and emissions. They selected the LN, Pearson type V, and Weibull distributions for fitting these data. Results showed that the LN distribution is a better representation of PM2.5 and PM10	Lu and Fang (2002)
NO_2, PM10, and SO_2	Shanghai, China June 1, 2000, to May 31, 2003	The EV, gamma, LN, and Pearson type V distributions were used in this study. The authors concluded that the best fit to NO_2, PM10, and SO_2 data was obtained with the EV, LN, and Pearson type V distributions, respectively	Kan and Chen (2004)
PM10	Kuala Lumpur, Malaysia 1998–2002	The authors used the LN distribution to fit these data	Sedek et al. (2006)
CO	Two air quality control stations in Delhi, India 1997–1999.	The authors found that this data set exceeds the Indian air quality guidelines during winter time and concluded that most of the data are fitted well with the LL[n] distribution, except for one data set, in which the LN distribution was more adequate. The reason for this discrepancy was the absence of high concentrations for those data in which the LN model was more appropriate. This is reasonable since the LL model has heavier tails than the LN model, which probably provided a better fit in case of high concentrations	Gokhale and Khare (2007)
O_3	New York May–September, 1973	The author developed a truncated version of the inverted beta distribution and applied it to O_3 data	Nadarajah (2008)
CO, NO_2, and PM	Monitoring stations in Royapuram, Chennai, India February 2005 to December 2008	The authors used the chi-square, exponential, gamma, LN, normal, and Weibull distributions to fit these data. Results showed that the LN distribution fitted the data better	Deepa and Shiva (2010)

[a] CO = carbon monoxide
[b] NO_2 = nitrogen dioxide
[c] NO_x = nitrogen oxides
[d] O_3 = tropospheric ozone
[e] SO_2 = sulfur dioxide
[f] LN = lognormal
[g] EV = extreme value
[h] PM = particulate matter
[i] JSB = Johnson SB
[j] PM2.5 = particulate matter with diameter of 2.5 μm or less
[k] PM10 = particulate matter with diameter of 10 μm or less
[l] PM8 = particulate matter with diameter of 8 μm or less
[m] GEV = generalized EV
[n] LL = log-logistic

3.3 Ott (1990)'s Physical Justification

Ott (1990) proposed a physical mechanism to explain why the LN distribution fits pollutant air concentrations well. Ott (1995) and Nevers (2000) observed that the fit of the LN distribution was better in the absence of extreme concentrations of pollutants.

3.4 Birnbaum–Saunders Air Contaminant Distribution (2008)

The BS distribution has theoretical arguments and suitable properties that are similar to those of the LN distribution, which allows it to be successfully applied for modeling air contamination data. Several versions of the BS distribution, useful for analyzing air pollutant data, have been developed. For example, Ferreira et al. (2012) proposed an EV version of the BS distribution and applied it for modeling extreme air pollutant concentrations. Details on studies in which the BS distributions have been used to model air pollutant concentrations are presented in Table 2.

4 Air Contamination in Santiago, Chile

Below, we describe the contamination faced by those living in Santiago, Chile, which, as mentioned above, is among the cities that have the highest air pollution levels in the world. We also address aspects of Chilean and international regulation of air quality.

4.1 Location

Santiago is located between 32–56° and 34–17° south (latitude) and between 69–47° and 71–43° west (longitude). The Metropolitan Region of Santiago is administratively divided into 6 provinces and 52 municipalities, covering an area of 15,554.5 km^2. The city is surrounded by both the Andes and coastal mountains, which create the Santiago Basin. The hills strongly restrict wind movement and air flow, limiting air renovation in the basin. Thus, in months of atmospheric stability, especially during the fall and winter months, air pollutants are trapped in the basin, which produces air contamination in the city of Santiago. During these periods, Santiago is exposed to episodes of sudden and critical rises in the levels of air contaminants, which are known as critical episodes, although they are usually short

Table 2 Details of air contaminant studies that were based on Birnbaum-Saunders models

Type of contaminant	Site and year of collected data	Comments on each statistical study	References
SO_2	Ten monitoring stations Santiago, Chile 2002	The authors concluded that generalized BS[a] distributions such as BS-Laplace, BS-logistic, and BS-t[b] showed better results and improved the precision of the exceedance probabilities	Leiva et al. (2008)
SO_2, O_3	SO_2 levels from Santiago, Chile (2002), and O_3 from New York 1973	The authors used the BS distribution to model two data sets. As in Leiva et al. (2008), the authors showed an improvement in the precision when BS distributions are used in place of LN distributions	Leiva et al. (2010)
SO_2	Monitoring stations in Santiago, Chile 2002	The authors developed an extended version of the BS distribution based on the skew distribution of Mudolkar-Hutson and compared it to the BS, IG[c], LN, length-biased BS, BS-slash, and BS-t distributions. The conclusion was that the extended model makes a better description than the other distributions	Vilca et al. (2010)
O_3	New York May–September, 1973	The authors developed a variation of the BS distribution, particularly useful for extreme concentrations. This distribution works very well with O_3 data	Vilca et al. (2011)
O_3	New York May–September, 1973	The authors developed an EV version of the BS distribution, appropriate for extreme concentrations. This distribution works much better than Vilca et al. (2011)'s extension and better than the EV distributions for O_3 data	Ferreira et al. (2012)

[a]BS = Birnbaum-Saunders
[b]BS-t = Birnbaum-Saunders-t Student
[c]IG = inverse Gaussian

lived. Critical episodes originate from coincidental meteorological factors, which prevent proper ventilation of the air of Santiago increasing its contamination (Rutllant and Garreaud 1995).

4.2 Wind and Thermal Inversion Effects

Bencala and Seinfeld (1976) indicated that wind velocity influences pollutant concentrations, specifically CO. Lu and Fang (2002) showed that the air contaminant concentration of Sha-Lu, Taiwan, varied according to local meteorological conditions and emission levels. These authors established a relationship between wind speed and the resulting level of air contamination; see also Leiva et al. (2011).

Thermal inversions in Santiago are critical and occur practically the whole year. In central Chile, such inversions predominantly result from high pressure systems and consist of air temperature increases at elevations of 700–1,000 m. This condition prevents air circulation and leads to contaminated air being trapped in the lower atmospheric layers. During the fall and winter months, an additional thermal inversion layer is generated from cooling of the earth's surface. Consequently, tropospheric air becomes very stable, when inversion layers of both types simultaneously occur, giving rise to very critical air contamination episodes (Garreaud and Rutllant 2004).

4.3 Monitoring Stations and Data Measurement

The automatic monitoring of air quality of the Metropolitan Region network (MACAM1—monitoreo automático de calidad del aire de la Región Metropolitana—in Spanish) was established in 1987, and it was the first monitoring program set up in Santiago. This network was funded by the Inter-American Development Bank. MACAM1 consisted of four stations located in downtown Santiago, and a fifth mobile station was situated farther east in the town of Las Condes. MACAM1 was renovated in 1997 with the addition of modern equipment, designed to provide the Chilean government with faster and more reliable reports of Santiago's air quality. New stations were installed and funded, mainly by the Japanese government. The MACAM2 network (successor monitoring network after renovation) is currently comprised eight fixed stations and one mobile station. The locations and designations of these stations are as follows: (S1) Providencia, (S2) Independencia, (S3) La Florida, (S4) Las Condes, (S5) Santiago, (S6) Pudahuel, (S7) Cerrillos, and (S8) El Bosque. The locations of these stations are shown in Fig. 1. The layout was designed so that the stations could measure and transmit average real concentrations of air pollutants that were not influenced by local sources of contamination. These stations were also located to meet geographical criteria, as they were designed to represent different population zones. Thus, the statistical data provided by these stations allow the Chilean authorities to have reliable information that can be employed to make useful decisions for the exposed population.

Stations S1 to S8 provide input on the concentrations of CO, NO_2, O_3, PM2.5, PM10, SO_2, and total hydrocarbons. PM data, for example, are collected as

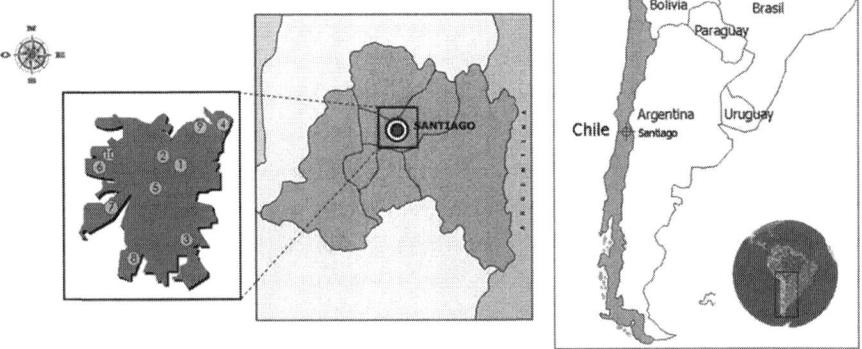

Fig. 1 Locations of air monitoring stations in Santiago (*left*), in the Metropolitan Region (*center*), and where they are located in Chile (*right*)

5-min-average bundles that are sent to a central computer in the Metropolitan Environmental Health Service of the Chilean government (SESMA, Servicio de Salud Metropolitano del Ambiente del gobierno de Chile, in Spanish) to comprise a 1-h analysis. In turn, the SESMA central computer distributes data to two additional computers: (1) in the National Environmental Center (CENMA, Centro Nacional del Medio Ambiente, in Spanish), www.cenma.cl, a foundation within the University of Chile, and (2) in the Ministry of the Environment of the Chilean government. A second semiautomatic monitoring network measures and analyses PM composition. The monitoring process requires that filters are exposed to air for 24 continuous hours. Results are obtained every 3 or 4 days. The objective of this process is to have a historical record for evaluating PM behavior over time. Daily calibration of the equipment is performed against defined atmospheric standards (artificial air).

4.4 Human Health Effects and International Guidelines

In Santiago, many patient visits to the emergency room have been epidemiologically associated with air pollution (Cakmak et al. 2009). Such patient visits result from acute bronchitis, lower airway disease in the elderly, obstructive bronchial syndrome, and pneumonia in children (Prieto et al. 2007; Muñoz and Carvalho 2009). Air contamination may also present a risk for epilepsy, headaches, and venous thromboembolic disease (Dales et al. 2009, 2010; Cakmak et al. 2010).

Chilean guidelines for air quality are established by the Ministry of the Environment and include safe levels for CO, NO_2, O_3, PM2.5, PM10, and SO_2. The safe limits of these contaminants are displayed in Table 3. Maximum concentrations (in $\mu g/m^3 N$), according to Chilean, American, and European guidelines, are provided in Table 4.

Table 3 Levels of environmental hazard associated with the indicated concentrations (in μg/m³N) according to Chilean guidelines

Levels	NO_2	SO_2	CO	O_3	PM2.5	PM10
Alert	1,130–2,259	1.962–2.615	17–33	400–799	80–109	195–239
Emergency prior	2,260–2,999	2.616–3.923	34–39	800–999	110–169	240–329
Emergency	≥3,000	≥3.924	≥40	≥1,000	≥170	≥330

NO_2, O_3, and SO_2 correspond to 1-h concentrations, CO to 8-h, and PM2.5 and PM10 to 24-h

Table 4 Maximum permitted guideline concentrations (in μg/m³N) according to Chilean, American, and European standards

	1-h			8-h			24-h			1-year		
Pollutant	Chile	EPA[a]	EEC[b]	Chile	EPA	EEC	Chile	EPA	EEC	Chile	EPA	EEC
CO	40	40	30	10	10	10	–[c]	–	–	–	–	–
NO_2	400	400	200	–	–	–	–	–	–	100	100	40
PM2.5	–	–	–	–	–	–	50	35	–	20	15	20
PM10	–	–	–	–	–	–	150	150	100	50	–	30
O_3	160	–	–	120	150	120	–	–	–	–	–	–
SO_2	–	–	–	–	–	–	250	365	125	80	80	50

[a]EPA = Environmental Protection Agency (US)
[b]EEC = European Economic Community
[c] "–" = missing values, which are not considered by the corresponding organization

4.5 A Study in Santiago, Chile

Gramsch et al. (2006) published a study of O_3 and PM concentrations in Santiago. These concentration results were delivered by MACAM2 during the year 2000. The aim of the study was to determine seasonal tendencies and spatial distribution of contaminating agents during the course of 1 year. Results clearly demonstrated concentration-season dependence, with O_3 and PM10 concentrations being higher in summer and winter months, respectively. The authors also noted that during the winter season, PM10 reached its highest level at night, but only in zones that had lower elevation. The reason for this is that wind entrapment and thermal inversions occur at lower (downtown) elevations. Contaminant levels in other zones (e.g., Las Condes) may not experience such daily PM fluctuations. The authors applied a cluster multivariate analysis to their O_3 and PM10 data, and defined four extended areas that displayed similar contamination patterns. These four areas were (A1) Providencia and Independencia, (A2) La Florida and El Bosque, (A3) Las Condes, and (A4) Santiago, Pudahuel, and Cerrillos. The fact that both contaminants had similar behavior is a clear indication that concentration levels are determined by meteorological and topographic conditions.

5 Application to PM10 Concentrations in Santiago, Chile

Below, we illustrate the application of a statistical methodology based on BS distributions to unpublished air contamination data from Santiago, Chile.

5.1 Problem Under Analysis and Data

Due to a combination of meteorological and topographic factors, Santiago, Chile, endures bad atmospheric ventilation during both winter and summer months. In winter, there is an accumulation of PM and gaseous contaminants, whereas increased solar radiation during the summer favors ozone-producing photochemical reactions. As mentioned in Sect. 4, CO, NO_2, PM10, and SO_2 are the main contributors to air quality problems in Santiago. In addition, NO_2 and SO_2 are precursors of PM10. Both groups of these pollutants were included in the PM10 data collected during 2003 by the Metropolitan Environmental Health Service. Therefore, we utilize these data for our analysis below, which are available at www.mma.gob.cl.

Chilean air quality guidelines indicate the maximum level at which contaminant concentrations become harmful to human health (see Sects. 2 and 4.4). At such levels, one can calculate statistical indicators such as percentiles and exceedance probabilities, i.e., by computing the probability that a concentration level exceeds a value established by the guidelines for a given period of time. Thus, our methodology consists of five steps. First, we conduct an autocorrelation analysis. Second, we carry out an exploratory data analysis (EDA) of the PM10 concentrations collected in 2003 at the monitoring stations in Santiago, Chile (see Fig. 1). Third, based on this EDA, we propose statistical distributions that are appropriate to model the PM10 data. We employ usual goodness-of-fit techniques to test for the most appropriate distribution for these data. Fourth, we utilize these distributions to calculate exceedance probabilities and percentiles for establishing administrative targets that may be useful to the Chilean government for establishing regulatory environmental alerts. Finally, we recommend how this information can be useful for preventing adverse effects on human health of the population of Santiago, Chile.

PM10 concentration data under analysis are (1) expressed as $\mu g/m^3 N$, (2) collected at monitoring stations Las Condes (S4) and El Bosque (S8), in Santiago during 2003, and (3) obtained as 1 h (hourly) average values. PM10 data are used for a 1-year analysis (2003) and then for a monthly analysis, mainly focused in April 2003 (see Sect. 4.3 and Gramsch et al. 2006). The numbers of annual and monthly observations under analysis are 8,760 and 720, for the year 2003 and the month of April 2003, respectively. Missing data were not considered in the analysis. For this illustration, we select data only from stations S4 and S8, mainly because the data from these stations are better conformed to situations of low and high stability, which then allow us to analyze for different pollution patterns.

Table 5 Descriptive summary of PM10 levels (in μg/m³N) for the indicated station for 2003

Station	Median	Mean	SD[a]	CV[b]	CS[c]	CK[d]	Range	Min.	Max.	n[e]
S4[f]	47.00	54.37	37.42	68.33%	1.44	6.82	397.00	1.00	398.00	8,727
S8[g]	68.00	83.18	62.16	74.73%	1.78	7.64	513.00	1.00	514.00	8,715

[a] SD = standard deviation
[b] CV = coefficient of variation
[c] CS = coefficient of skewness
[d] CK = coefficient of kurtosis
[e] n = sample size
[f] S4 = monitoring station "Las Condes"
[g] S8 = monitoring station "El Bosque"

5.2 Autocorrelation Analysis

Horowitz (1980) discussed certain conditions in which autocorrelation between successive air pollutant concentrations can be ignored for the purpose of estimating the statistical distribution of such concentrations. However, these concentrations are generally correlated and may be affected by random and systematic fluctuations due to changes in time and emission patterns. Nevertheless, it is possible to identify periods during the year in which concentration series are nearly non-mobile and uncorrelated. Assuming these conditions, several statistical distributions may be useful for performing air quality data analysis based on independent observations. For PM10 concentrations, we conduct an autocorrelation analysis that does not show a correlation between these concentrations. Therefore, our analysis agrees with those of Morel et al. (1999), Gokhale and Khare (2007), and Vilca et al. (2011).

5.3 Exploratory Data Analysis

As specified in Table 4, Chilean air quality guidelines allow a maximum PM10 24-h average concentration of 150 μg/m³N and a maximum annual concentration of 50 μg/m³N. In addition, this Chilean guideline establishes that PM annual series must have over 75% of the available data. We note that our data of PM10 complies with this requirement, which validates our conclusions by the Chilean air quality guidelines for PM10 (CONAMA 1998).

In Table 5, we provide a descriptive summary of PM10 data for stations S4 and S8. This summary includes the mean, median, standard deviation (SD), and coefficients of variation (CV), of skewness (CS), and of kurtosis (CK) of the analyzed data, among other indicators. Figs. 2, 3, and 4 display times series, histograms, and boxplots for PM10 data, respectively. From Table 5, we note that

1. The maximum concentration is exceeded in both stations (S4 and S8). This means that PM10 concentrations at both stations S4 and S8 far exceed 150 μg/m³N, but concentrations at S4 are less extreme than at S8.

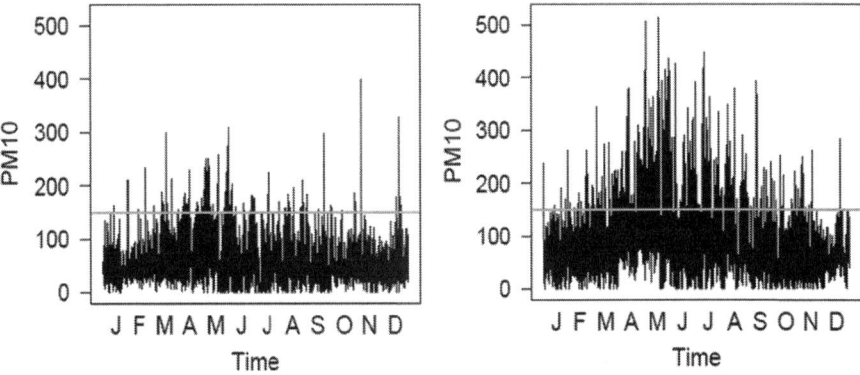

Fig. 2 Time series for PM10 levels measured at S4 (*left*) and S8 (*right*). S4 = monitoring station "Las Condes", S8 = monitoring station "El Bosque"

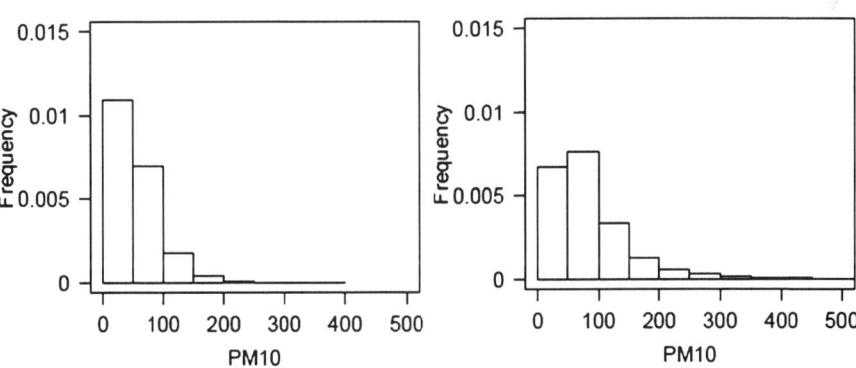

Fig. 3 Histograms for PM10 levels measured at S4 (*left*) and S8 (*right*)

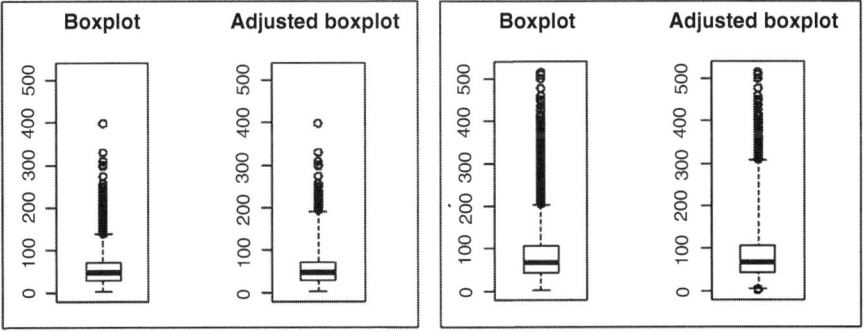

Fig. 4 Usual and adjusted boxplots for PM10 levels measured at S4 (*left*) and S8 (*right*)

2. The CS and CK of PM10 data at both S4 and S8 show positively skewed distributions with a high kurtosis.

From Figs. 2, 3, and 4, we note that

1. Time series for PM10 levels in Fig. 2 show a gray line representing the allowed maximum concentration. In the S8 time series, there is an important rise in fall and winter months, while concentrations at S4 are more stable than at S8. This may be due to the higher elevation of S4 than S8.
2. Histograms of PM10 data are shown in Fig. 3, and provide evidence for positive asymmetry—skewness—(CS>0) of the statistical distribution of the data and heavy tails (CK>3).
3. The boxplots of PM10 data in Fig. 4 show a huge number of possibly atypical data in both stations. It is worth noting that Tukey (1977) originally developed the boxplot to determine whether data have a normal distribution that is symmetrical or not. However, when contaminant concentration data have an asymmetric distribution, the usual boxplot may result in observations being classified as atypical, despite these are not atypical. Hubert and Vandervieren (2008) produced an adjusted boxplot for use with asymmetric data. In Fig. 4, we show both the usual and adjusted boxplots for PM10 data measured at S4 and S8. These plots show how data that are considered to be atypical, when depicted in the usual boxplot, become non-atypical by the adjusted boxplot.

Table 6 provides a descriptive summary of the monthly PM10 data for stations S4 and S8. This summary is coherent with the annual analysis and with the graphical analysis.

Figure 5 shows adjusted boxplots of monthly PM10 levels for S4 and S8. In this figure, we note that both stations in the months of May produce higher PM10 concentrations, but fewer atypical observations. In Fig. 5 (left boxplots), we note that January and October were months that had a high number of atypical PM10 concentrations for S4. However, in Fig. 5 (right boxplots), months with high concentrations and few atypical data for S8 were February and November. High variability, positive skewness, and high kurtosis were detected for nearly every month and station. Based on the results from the EDA of PM10 data, GBS distributions seem to be very good options for modeling PM10 concentrations in Santiago. These distributions are good options because they allow for an appropriate accommodation of data variability, skewness, and kurtosis, due to their flexibility and to the production of robust parameter estimates in the presence of atypical data, as is the case of PM10 data.

5.4 Selection and Validation of the Model

The β_1–β_2 chart is a well-known tool to simultaneously fit several data sets. This chart is based on standardized statistical moments (expected values) such as the CS (viz. β_1) and CK (viz. β_2). For more details about these charts, see Tsukatani and Shigemitsu (1980) and Leiva et al. (2008).

Table 6 Descriptive summary of PM10 levels (in μg/m³N) for the indicated station for 2003 by month

Month	Median	Mean	SD	CV	CS	CK	Range	Min.	Max.	n	Percentage of missing data
S4											
January	40.00	44.85	29.36	65.47%	1.73	7.80	209.00	1.00	210.00	740	0.54
February	44.00	47.99	22.41	46.69%	1.67	10.51	226.00	7.00	233.00	672	0.00
March	56.50	62.59	33.08	52.92%	1.42	7.37	297.00	3.00	300.00	744	0.00
April	66.00	71.72	39.30	54.80%	0.64	3.06	229.00	1.00	230.00	717	0.42
May	68.00	76.76	57.05	74.32%	0.83	3.45	307.00	1.00	308.00	743	0.13
June	43.00	53.27	41.06	77.08%	0.94	3.45	203.00	1.00	204.00	717	0.42
July	45.00	51.37	39.13	76.16%	0.71	3.09	222.00	1.00	223.00	744	0.00
August	50.00	57.79	39.87	68.99%	0.84	3.34	208.00	1.00	209.00	743	0.13
September	37.00	43.25	31.79	73.51%	1.61	9.35	296.00	1.00	297.00	714	0.83
October	45.00	48.85	25.32	51.83%	1.27	5.90	184.00	2.00	186.00	744	0.00
November	40.00	55.88	28.37	63.20%	3.41	36.66	397.00	1.00	398.00	710	1.39
December	43.00	48.25	30.11	62.42%	2.16	15.40	328.00	1.00	329.00	739	0.67
S8											
January	54.00	61.95	35.91	57.96%	1.34	6.35	260.00	1.00	261.00	719	3.36
February	66.00	71.65	38.29	53.44%	1.23	5.87	262.00	1.00	263.00	672	0.00
March	74.00	82.26	44.67	54.30%	1.12	5.42	243.00	1.00	344.00	742	0.27
April	100.00	113.61	57.34	50.47%	1.39	5.24	362.00	17.00	379.00	720	0.00
May	129.50	147.78	93.24	63.09%	0.89	3.76	513.00	1.00	514.00	744	0.00
June	79.00	93.44	71.22	76.22%	0.98	3.78	425.00	1.00	426.00	719	1.00
July	78.00	97.58	77.95	79.88%	1.28	4.81	446.00	1.00	447.00	742	0.14
August	77.00	92.42	62.93	68.09%	1.32	5.19	379.00	1.00	380.00	744	0.00
September	52.00	62.00	47.93	77.31%	1.81	9.24	392.00	1.00	393.00	719	0.14
October	45.00	48.85	25.32	51.83%	1.27	5.90	184.00	2.00	186.00	744	0.00
November	48.00	53.72	35.23	65.58%	1.56	7.84	262.00	1.00	263.00	713	0.97
December	56.00	59.79	28.10	46.99%	1.33	8.26	281.00	1.00	282.00	737	0.94

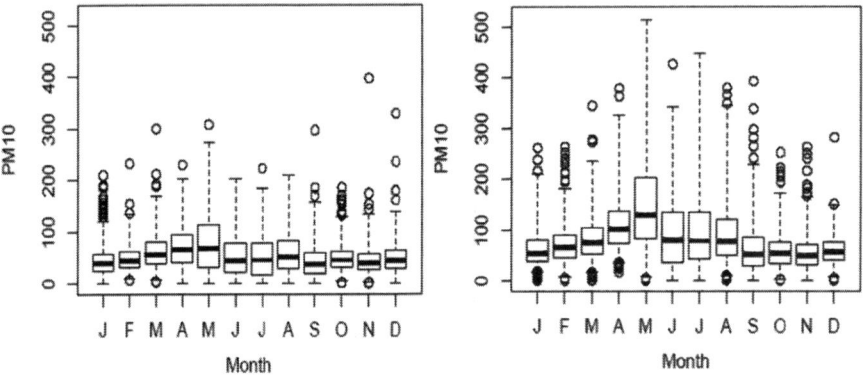

Fig. 5 Adjusted boxplots for monthly PM10 levels measured at S4 (*left*) and S8 (*right*)

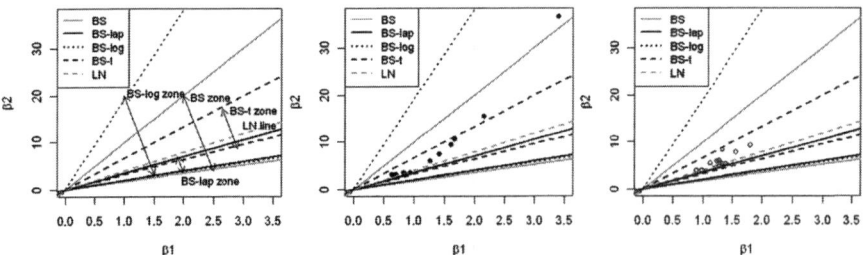

Fig. 6 β_1–β_2 charts with fit zone (*left*) and individual cases for PM10 levels measured at S4 (*center*) and S8 (*right*)

Figure 6 (left) shows β_1–β_2 charts with different fit zones for some GBS distributions, such as the BS, BS-Laplace, BS-logistic, and BS-Student-t (BS-t) distributions. Interestingly, by utilizing this method (not shown here), data from monitoring stations S1 to S8 are approximately grouped in a way that is similar to the cluster analysis carried by Gramsch et al. (2006), which was discussed in Sect. 4.5. The β_1–β_2 chart for all the stations (not shown here) indicates that the GBS distributions are globally appropriate for PM10 data, and these are, in general, more appropriate for modeling PM10 data than the LN distribution, which is habitually used to model air contamination data. This situation is expected, primarily for S4, because its proportion of atypical data was high. Figure 6 (center) and (right) displays β_1–β_2 charts for PM10 data in S4 (bold dots) and S8 (white dots), respectively. The figure shows that the data can be well approached by the GBS family of statistical distributions. Specifically, at S4, the chart shows that the BS-logistic (dotted bold line) and BS-t (segmented bold line) distributions are suitable, mainly in the more unstable months, and, at S8, BS-t (segmented bold line) and LN (segmented gray line) distributions are appropriate. The reason for this conclusion is because the points representing each sample by their sample CS and CK are more concentrated in the GBS zones

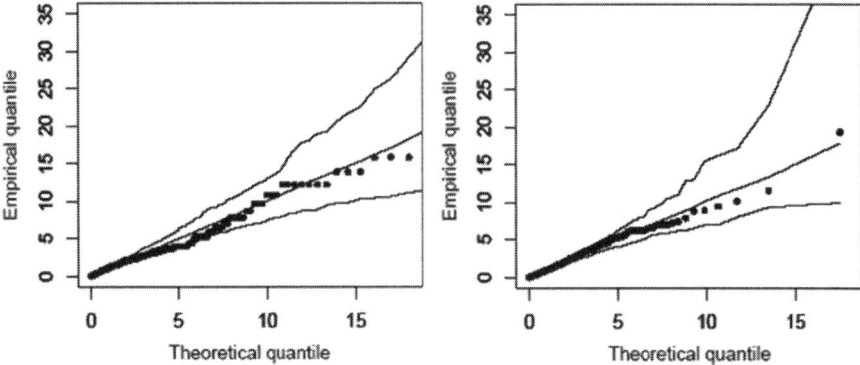

Fig. 7 Probability plots with envelopes for PM10 levels measured at S4 (*left*) and S8 (*right*), using the BS-t distribution

than in the LN zone, mainly at S4. This approach demonstrates the necessity for modeling PM10 data by means of a heavy-tailed air contaminant distribution. In summary, we postulate that, in most cases, the actual distribution of PM10 data can best be approached by using GBS distributions. We propose the BS-t distribution because it provides robust parameter estimates in the presence of atypical concentrations, which does not occur with the LN distribution.

As mentioned by Leiva et al. (2008), the charts in Fig. 6 are useful for considering or discarding distributions. These charts can be used as an initial method to test goodness of fit. In cases where more than one distribution is selected by using these charts, it is appropriate to consider further goodness-of-fit techniques. These techniques allow us to obtain the final parametric statistical distribution that can be used for making administrative decisions. To confirm the results obtained by goodness-of-fit methods that are based on moments for PM10 data, we use a probability plot that includes bands (envelope) facilitating the visualization of the fit. Figures 7 and 8 display probability plots with envelopes for PM10 data, based on the BS-t and LN distributions of data collected at S4 and S8 during April 2003. These plots show that there is good agreement between the BS-t distribution and the data. For S4, several points located in the right tail lie outside of the LN probability plot with envelope. This indicates that a heavy-tailed distribution is needed for analyzing these PM10 data. Such an aspect is corroborated by means of the BS-t probability plot with envelope, where all points lie inside of this envelope. For S8, however, both distributions (BS-t and LN) show a good fit to PM10 data.

5.5 Computing Exceedance Probabilities and Percentiles

Administrative targets to control air pollution usually belong to the [98.0–99.9]th percentile range. Thus, we need to closely determine the distribution in the range of the higher contaminant concentrations.

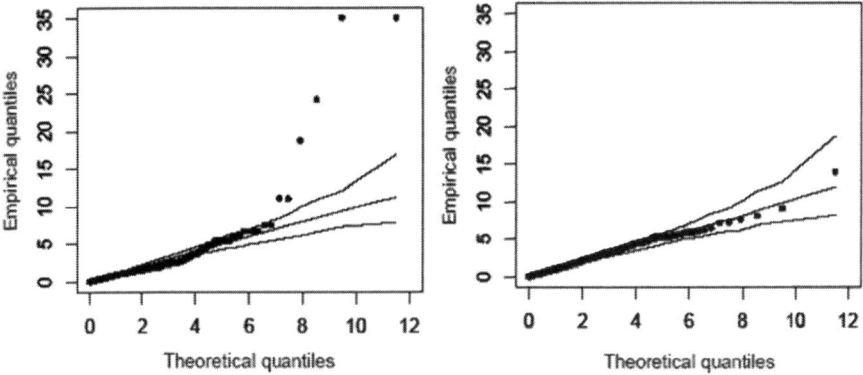

Fig. 8 Probability plots with envelopes for PM10 levels measured at S4 (*left*) and S8 (*right*), using the LN distribution

Table 7 Estimated 98th percentile for PM10 (in μg/m³N) at the indicated stations and using the indicated model

Station	BS	LN	BS-t
S4	242.44	245.20	311.99
S8	268.30	270.96	272.88

For air contamination in Santiago, we use the well-known maximum likelihood method for estimating the parameters of the BS, BS-t, and LN distributions that are based on PM10 levels recorded for S4 and S8 during April 2003. From these estimates, we compute the probabilities for detecting levels greater than 150 μg/m³N, i.e., P(PM10>150), according to the Chilean air quality guidelines given in Table 4 as 24-h average PM10 concentration. These exceedance probabilities are 0.0899, 0.0908, and 0.0956 for S4, and 0.21150, 0.20631, and 0.20231 for S8, using the LN, BS, and BS-t distributions, respectively. This indicates that, for S4, both distributions (BS and LN) are less capable of detecting higher pollution levels because their probabilities are smaller, as expected. However, as also expected, there is a different situation at S8, where the levels are more stable. Therefore, if data from station S4 are used to establish the administrative target in Santiago, the LN distribution is not a suitable model because high levels of contaminants could not be detected by this distribution and an erroneous decision could be made. Administrative targets to abate air pollution belonging to the [98.0–99.9]th percentile range are presented in Table 7 for the indicated distributions. From this table, we note that, by using the BS-t distribution, extreme concentrations could be detected, mainly for S4.

5.6 Connecting Statistical Information with Adverse Effects on Human Health

As mentioned, the adverse effects of PM10 exposure vary with its composition, and may occur only after extended exposure. Official environmental guidelines use air pollutant concentration distribution to compute exceedance probabilities and percentiles that allow for determining administrative regulatory targets or establishing environmental alerts. These alerts benefit human health because they influence governmental action to reduce air pollutant emissions. Thus, it is necessary to precisely establish the distribution of air pollutant concentrations in the range of the higher contaminant concentrations. For example, air quality regulatory guideline of the Chilean government indicates that the 98th percentile of the PM10 24-h concentration collected at any monitoring station cannot exceed 150 $\mu g/m^3 N$. The Chilean guideline also indicates that chronic health effects are those produced by the action of PM10 concentrations during prolonged periods of exposure, and which are manifested by an increment of the incidence and severity of diseases such as asthma, bronchitis, and emphysema; see Chilean air quality regulatory guideline (CONAMA 1998).

6 Conclusions

In this chapter, we have provided information about air contaminants and their adverse health effects on humans. We have also reviewed the statistical distributions that have been used to model air quality and described the problem of air contamination in Santiago, Chile, which can be extended to other geographical sites. The following sentences are synoptic conclusive statements designed both to convey the authors' objectives in writing this work and to facilitate the understanding of data and results presented in the previous paragraphs. In this chapter, we have undertaken to

1. Establish an epidemiological surveillance to qualitatively and quantitatively determine the effects of contaminants on human health during different periods of the year. For example, the Chilean guideline states that chronic health effects are produced by prolonged exposure to PM10 and are manifested by an increment of the incidence and severity of diseases such as asthma, bronchitis, and emphysema (CONAMA 1998),
2. Identify the pollutants that pose higher risk for damaging human health.
3. Detect concentrations of pollutants that cause adverse health effects in humans.
4. Choose reliable pollutant quantification systems consisting of monitoring stations located at representative residential areas.
5. Propose a representative and reliable strategy for pollutant concentration data management. Such a strategy should be able to establish

    (a) What pollutant is suitable for a prediction methodology (e.g., PM10)
    (b) What monthly concentration should be considered (e.g., 24-h average in each monitoring station per month)
    (c) What percentage of the programmed measurements for each month must be considered to validate the data of such a month (e.g., at least 75%)
    (d) When to begin the analysis period (generally from January 1st to December 31st of the same year for annual data and from the first day in a month to the previous day of the start of the following month, for monthly data)

6. Have guidelines that adequately protect human health from a given pollutant, based on levels that allow administrative decision making. For example, declaring an environmental alert of PM contamination when at least one air quality monitoring station detects a level above 195 and below 240, or to declare a preemergency situation when a level of 330 is exceeded in at least one station. Chilean air quality regulatory guideline indicates that a value for PM10 greater than 150 µg/m³N as a 24-h concentration is not allowed. Specifically, this guideline is surpassed when the 98th percentile of the PM 24-h concentration, collected in any monitoring station, exceeds 150 µg/m³N (CONAMA 1998).
7. Have a statistical methodology for administrative decision making such as that proposed in Sect. 5 of this chapter. This methodology should consist in

   (a) Setting up a database that allows for representative and reliable pollutant concentration data management.
   (b) Conducting an autocorrelation analysis so that space–time component can be discarded or not, and then the methodology to be used considers this component of dependence or otherwise the data are analyzed as a random sample (independence).
   (c) Carrying out exploratory data analysis to obtain information about the most appropriate statistical distribution for modeling the data.
   (d) Conducting a selection and validation study for deciding the best model to be used for administrative decision making.
   (e) Applying the selected and validated model to obtain relevant information based on official guidelines. For example, to estimate whether the 98th percentile of the PM 24-h concentration, collected in any monitoring station, is greater than 150 µg/m³N.

8. Set up a decontamination plan for mitigating the risk of adverse health effects and to have a plan for future episodes of critical contamination.
9. Propose a predictive system for managing critical contamination episodes that allow for timely and effective human health effect protection.
10. Consider that any guideline exceedance declaration should not result from its first exceedance. For example, the guideline is considered to be surpassed if measurements over seven or more days produce results that exceed 150 µg/m³N before the end of the first annual measurement period certified by a competent health institution.

7 Summary

The use of statistical distributions to predict air quality is valuable for determining the impact of air chemical contaminants on human health. Concentrations of air pollutants are treated as random variables that can be modeled by a statistical distribution that is positively skewed and starts from zero. The type of distribution selected for analyzing air pollution data and its associated parameters depend on factors such as emission source and local meteorology and topography. International environmental guidelines use appropriate distributions to compute exceedance probabilities and percentiles for setting administrative targets and issuing environmental alerts. The log-normal distribution is frequently used to model air-pollutant data. This distribution bears a relationship to the normal distribution, and there are theoretical- and physical-based mechanistic arguments that support its use when analyzing air-pollutant data. Other distributions have also been used to model air pollution data, such as the beta, exponential, gamma, Johnson, log-logistic, Pearson, and Weibull distributions. One model also developed from physical-mechanistic considerations that has received considerable interest in recent years is the Birnbaum–Saunders distribution. This distribution has theoretical arguments and properties similar to those of the log-normal distribution, which renders it useful for modeling air contamination data. In this review, we have addressed the range of common atmospheric contaminants and the health effects they cause. We have also reviewed the statistical distributions that have been used to model air quality, after which we have detailed the problem of air contamination in Santiago, Chile. We have illustrated a methodology that is based on the Birnbaum–Saunders distributions to analyze air contamination data from Santiago, Chile. Finally, in the conclusions, we have provided a list of synoptic statements designed to help readers understand the significance of air pollution in Chile, and in Santiago, in particular, but that can be useful to other cities and countries.

Acknowledgments The authors wish to thank the editor, Dr. David M. Whitacre, and the referees for their constructive comments on an earlier version of this chapter, which resulted in the current version. C. Marchant gratefully acknowledges support from the scholarship "President of the Republic" of the Chilean government of which she was a recipient during her studies in engineering in statistics in the University of Valparaiso which concluded with this work. The research of V. Leiva was partially supported by FONDECYT 1120879 grant from the Chilean government. The research of A. Sanhueza was partially supported by FONDECYT 1080409.

References

Balakrishnan N, Leiva V, Sanhueza A, Vilca F (2009) Estimation in the Birnbaum-Saunders distribution based on scale-mixture of normals and the EM-algorithm. SORT 33:171–192

Barry PJ (1971) Use of Argon-41 to study the dispersion of stack effluents. Proc Symp Nucl Techn Environ Pollut Int. Atomic Energy Agenc 241–253

Bencala KE, Seinfeld JH (1976) On frequency distribution of air pollutant concentrations. Atmos Environ 10:941–950

Berger A, Melice JL, Demuth CL (1982) Statistical distribution of daily and high atmospheric SO2 concentration. Atmos Environ 16:2863–2877

Birnbaum ZW, Saunders SC (1969) A new family of life distributions. J Appl Probab 6:319–327

Brauer M, Hoek G, van Vliet P, Meliefste K, Fischer P, Gehring U, Heinrich J, Cyrys J, Bellander T, Lewne M, Brunekreef B (2003) Estimating long-term average particulate air pollution concentrations, application of traffic indicators and geographic information systems. Epidemiology 14:228–239

Brook RD, Franklin B, Cascio W, Hong Y, Howard G, Lipsett M, Luepker R, Mittleman M, Samet J, Smith SC Jr, Tager I (2004) Expert panel on population and prevention science of the American Heart Association. Air pollution and cardiovascular disease: a statement for healthcare professionals from the Expert Panel on Population and Prevention Science of the American Heart Association. Circulation 109:2655–2671

Brook RD, Rajagopalan S, Arden C, Brook JR, Bhatnagar A, Diez-Roux A, Holguin F, Hong Y, Luepker R, Mittleman MA, Peters A, Siscovick D, Smith S, Whitsel L, Kaufman J (2010) Particulate matter air pollution and cardiovascular disease: an update to the scientific statement from the American Heart Association. Circulation 121:2331–2378

Cakmak S, Dales RE, Gultekin T, Vidal CB, Farnendaz M, Rubio MA, Oyola P (2009) Components of particulate air pollution and emergency department visits in Chile. Arch Environ Occup Health 64:148–155

Cakmak S, Dales RE, Vidal CB (2010) Air pollution and hospitalization for epilepsy in Chile. Environ Int 36:501–505

CEPAL (2000) The equity gap: a second assessment. Second Regional Conference in Follow-up to the World Summit for Social Development, Santiago, Chile

Cicolella A (2008) Les composes organiques volatils (COV): definition, classification et proprietes. Rev Mal Respir 25:155–163

Cifuentes L, Borja-Aburto VH, Gouveia N, Thurston G, Davis DL (2001) Assessing health benefits of urban air pollution reductions associated with climate change mitigation (2000–2020): Santiago, Sao Paulo, Mexico City, and New York City. Environ Health Perspect 109:419–425

Cohen AJ, Anderson HR, Ostro B et al (2004) Urban air pollution. In: Ezzati M, Lopez AD, Rodgers A, Murray CJL (eds) Comparative quantification of health risks: global and regional burden of disease attributable to selected major risk factors. World Health Organization, Geneva, pp 1353–1433

CONAMA (1998) Establece norma de calidad primaria para material particulado respirable PM10, en especial de los valores que definen situaciones de emergencia. Decreto 59 Gobierno de Chile, Santiago, Chile

Curran TC, Frank NH (1975) Assessing the validity of the lognormal model when predicting maximum air pollutant concentrations. Proc 68th Ann Meet Air Pollut Control Assoc 3:51–75

Dales RE, Cakmak S, Vidal CB (2009) Air pollution and hospitalization for headache in Chile. Am J Epidemiol 170:1057–1066

Dales RE, Cakmak S, Vidal CB (2010) Air pollution and hospitalization for venous thromboembolic disease in Chile. J Thromb Haemost 8:669–674

Deepa A, Shiva SM (2010) Statistical distribution models for urban air quality management. Int J Adv Geosci 16:285–297

Díaz-García JA, Leiva V (2005) A new family of life distributions based on elliptically contoured distributions. J Stat Plan Inference 128:445–457, Erratum: J Stat Plan Inference 137:1512–1513

Ferreira M, Gomes MI, Leiva V (2012) On an extreme value version of the Birnbaum-Saunders distribution. RevStat Stat J 10(2):181–210

Franco R, Sánchez-Olea R, Reyes-Reyes EM, Panayiotidis MI (2009) Environmental toxicity, oxidative stress and apoptosis: Ménage a trois. Mutat Res 674:3–22

Garreaud RD, Rutllant J (2004) Factores meteorológicos de la contaminación atmosférica en Santiago in episodios críticos de contaminación atmosférica en Santiago. R. Morales Colección de Química Ambiental, Universidad de Chile, Santiago de Chile, pp 9–36

Gifford FA (1974) The form of the frequency distribution of air pollution concentrations. Proc Symp Statistical Aspects of Air Quality Data, EPA-650/4-74-038, pp 3.1–3.7

Gokhale S, Khare M (2007) Statistical behavior of carbon monoxide from vehicular exhausts in urban environments. Environ Modell Softw 22:526–535

Gramsch E, Cereceda-Balic F, Oyola P, von Baer D (2006) Examination of pollution trends in Santiago de Chile with cluster analysis of PM10 and ozone data. Atmos Environ 40:5464–5475

Hedley AJ, Wong CM, Thach TQ, Ma S, Lam TH, Anderson HR (2002) Cardiorespiratory and all-cause mortality after restrictions on sulfur content of fuel in Hong Kong: an intervention study. Lancet 360:1646–1652

Hesterberg TW, Bunn WB, McClellan RO, Hamade AK, Long CM, Valberg PA (2009) Critical review of the human data on short-term nitrogen dioxide (NO_2) exposures: evidence for NO_2 no-effect levels. Crit Rev Toxicol 39:743–781

Holland DM, Terence FS (1982) Fitting statistical distributions to air quality data by the maximum likelihood method. Atmos Environ 16:1071–1076

Horowitz J (1980) Extreme values from a nonstationary stochastic process: an application to air quality analysis. Technometrics 22:469–478

Hubert M, Vandervieren E (2008) An adjusted boxplot for skewed distributions. Comp Stat Data Anal 52:5186–5201

Järup L (2003) Hazards of heavy metal contamination. Br Med Bull 68:167–182

Johnson NL, Kotz S, Balakrishnan N (1995) Continuous Univariate Distributions–Vol 2. Wiley, New York

Kampa M, Castanas E (2008) Human health effects of air pollution. Environ Pollut 151:362–367

Kan H, Wong CM, Vichit-Vadakan N, Qian Z (PAPA Project Team) (2010) Short-term association between sulfur dioxide and daily mortality: the public health and air pollution in Asia (PAPA) study. Environ Res 110:258–264

Kan H, Chen B (2004) Statistical distributions of ambient air pollutants in Shanghai, China. Biomed Environ Sci 17:366–372

Katsouyanni K (2003) Ambient air pollution and health. Br Med Bull 68:143–156

Kinney PL (2008) Climate change, air quality, and human health. Am J Prev Med 35:459–467

Larsen R (1971) A mathematical model for relating air quality measurements to air quality standards. Air Pollution Series, EPA-AP89

Latza U, Gerdes S, Baur X (2009) Effects of nitrogen dioxide on human health: systematic review of experimental and epidemiological studies conducted between 2002 and 2006. Int J Hyg Environ Health 212:271–287

Leiva V, Barros M, Paula GA, Galea M (2007) Influence diagnostics in log-Birnbaum-Saunders regression models with censored data. Comp Stat Data Anal 51:5694–5707

Leiva V, Barros M, Paula GA, Sanhueza A (2008) Generalized Birnbaum-Saunders distributions applied to air pollutant concentration. Environmetrics 19:235–249

Leiva V, Sanhueza A, Kelmansky S, Martinez E (2009) On the glog-normal distribution and its association with the gene expression problem. Comp Stat Data Anal 53:1613–1621

Leiva V, Vilca F, Balakrishnan N, Sanhueza A (2010) A skewed sinh-normal distribution and its properties and application to air pollution. Commun Stat Theory Methods 39:426–443

Leiva V, Athayde E, Azevedo C, Marchant C (2011) Modeling wind energy flux by a Birnbaum-Saunders distribution with unknown shift parameter. J Appl Stat 38:2819–2838

Listorti JA (1999) Is environmental health really a part of economic development—or only an afterthought? Environ Urban 11:89–100

Liu G, Niu Z, Van Niekerk D, Xue J, Zheng L (2008) Polycyclic aromatic hydrocarbons (PAHs) from coal combustion: emissions, analysis, and toxicology. Rev Environ Contam Toxicol 192:1–28

Lu H-C, Fang G-C (2002) Estimating the frequency distributions of PM10 and PM2.5 by the statistics of wind speed at Sha-Lu, Taiwan. Sci Total Environ 298:119–130

Lynn DA (1974) Fitting curves to urban suspended particulate data. Statistical Aspects of Air Quality Data, EPA 650/4-74-038, pp. 13.1–13.28

Maggiora CD, Lopez-Silva JA (2006) Vulnerability to air pollution in Latin America and the Caribbean Region. The World Bank, Latin America and the Caribbean Region, Environmentally and Socially Sustainable Development Department, Working paper No. 28

McConnell R, Berhane K, Gilliland F, Molitor J, Thomas D, Lurmann F, Avol E, Gauderman WJ, Peters JM (2003) Prospective study of air pollution and Bronchitic symptoms in children with asthma. Am J Respir Crit Care Med 168:790–797

Morel B, Yeh S, Cifuentes L (1999) Statistical distributions for air pollution applied to the study of the particulate problem in Santiago. Atmos Environ 33:2575–2585

Muñoz F, Carvalho MS (2009) Efecto del tiempo de exposición a PM10 en las urgencias por bronquitis aguda. Cad Saude Publica 25:529–539

Nadarajah S (2008) A truncated inverted beta distribution with application to air pollution data. Stoch Environ Res Risk Assess 22:285–289

Nevers D (2000) Air Pollution Control Engineering. McGraw-Hill, New York

Nuvolone D, Balzi D, Chini M, Scala D, Giovannini F, Barchielli A (2011) Short-term association between ambient air pollution and risk of hospitalization for acute myocardial infarction: results of the cardiovascular risk and air pollution in Tuscany (RISCAT) study. Am J Epidemiol 174:63–71

OECD-DAC (2000) Shaping the urban environment in the 21st century: from understanding to action, a DAC reference manual on urban environmental policy. Organization for Economic Cooperation and Development, Paris

Ostro P (2003) Air pollution and its impacts on health in Santiago, Chile. In: McGranahan G, Murray F (eds) Air Pollution And Health In Rapidly Developing Countries. Earthscan Publications Ltd., London

Ott WR (1990) A physical explanation of the lognormality of pollution concentrations. J Air Waste Manag Assoc 40:1378–1383

Ott WR (1995) Environmental Statistics and Data Analysis. Lewis Publishers, Boca Raton, FL

Ott W, Mage D (1976) A general-purpose univariate probability model for environmental data analysis. Comput Oper Res 3:209–216

Pani B (2007) Textbook of Environmental Chemistry. Jk International Publishing House, New Delhi

Pollack, R (1975) Studies of pollutant concentration frequency distributions. Environmental Monitoring Series, EPA-650/4-75-004

Prieto C, Mancilla FP, Astudillo OP, Reyes PA, Román AO (2007) Excess respiratory diseases in children and elderly people in a community of Santiago with high particulate air pollution. Rev Med Chil 135:221–228

Rumburg B, Alldredge R, Claiborn C (2001) Statistical distributions of particulate matter and the error associated with sampling frequency. Atmos Environ 35:2907–2920

Rutllant J, Garreaud R (1995) Meteorological air pollution potential for Santiago, Chile: towards an objective episode forecasting. Environ Monit Assess 34:223–244

Sanhueza A, Leiva V, Balakrishnan N (2008) The generalized Birnbaum-Saunders distribution and its theory, methodology and application. Commun Stat Theory Methods 37:645–670

Satterthwaite D (1997) Sustainable cities or cities that contribute to sustainable development. Urban Studies 34:1667–1691

Schecter A, Birnbaum LS, Ryan J, Constable J (2006) Dioxins: an overview. Environ Res 101: 419–428

Scriven RA (1971) Use of Argon-41 to study the dispersion of stack effluents. Proc Symp Nucl Techn Environ Pollut 253-255

Sedek JNM, Ramli NA, Yahaya AS (2006) Air quality predictions using lognormal distribution functions of particulate matter in Kuala Lumpur Malaysia. J Environ Manag 7:33–41

Simpson RW, Butt J, Jakeman AJ (1984) An averaging time model of SO2 frequency distributions from a single point source. Atmos Environ 18:1115–1123

Singpurwalla N (1972) Extreme values from a lognormal law with applications to air pollution problems. Technometrics 14:703–711

Soliman ASM, Palmer GM, Jacko RB (2006) Development of an empirical model to estimate real world fine particulate matter emission factors, the traffic air quality model. J Air Waste Manag Assoc 56:1540–1549

Taylor RA, Jakeman AJ, Simpson RW (1986) Modeling distributions of air pollutant concentrations. Identification of statistical models. Atmos Environ 20:1781–1789

Tsukatani T, Shigemitsu K (1980) Simplified Pearson distributions applied to air pollutant concentration. Atmos Environ 14:245–253

Tukey JW (1977) Exploratory Data Analysis. Addison-Wesley, Reading, MA

van Roosbroeck S, Wichmann J, Janssen NA, Hoek G, van Wijnen JH, Lebret E, Brunekreef B (2006) Long-term personal exposure to traffic-related air pollution among school children, a validation study. Sci Total Environ 368:565–573

Vilca F, Leiva V (2006) A new fatigue life model based on the family of skew elliptic distributions. Commun Stat Theory Methods 35:229–244

Vilca F, Sanhueza A, Leiva V, Christakos G (2010) An extended Birnbaum-Saunders model and its application in the study of environmental quality in Santiago, Chile. Stoch Environ Res Risk Assess 24:771–782

Vilca F, Santana L, Leiva V, Balakrishnan N (2011) Estimation of extreme percentiles in Birnbaum-Saunders distributions. Comp Stat Data Anal 55:1665–1678

White SS, Birnbaum LS (2009) An overview of the effects of dioxins and dioxin-like compounds on vertebrates, as documented in human and ecological epidemiology. J Environ Sci Health C 27:197–211

WHO –World Health Organization– (2005) Air quality management global update particulate matter, ozone, nitrogen oxide and sulfur dioxide. Reg Off Europe, Copenhagen

World Bank (2001) World development report 200012001, attacking poverty. Oxford University Press, Oxford

Advances in the Application of Plant Growth-Promoting Rhizobacteria in Phytoremediation of Heavy Metals

Hamid Iqbal Tak, Faheem Ahmad, and Olubukola Oluranti Babalola

Contents

1 Introduction 33
2 Phytoremediation 36
    2.1 Biological Availability of Metals in Soil 37
    2.2 Plant Uptake and Transport of Metals 38
    2.3 Plant Mechanisms for Metal Detoxification 39
3 Plant Growth-Promoting Bacteria 40
    3.1 How Do PGPR Combat Heavy-Metal Stress? 42
    3.2 Synergistic Interaction of PGPR and Plants in Heavy-Metal Remediation ... 43
    3.3 ACC Deaminase and Plant Stress Reduction from Ethylene 44
4 Summary 46
References 47

1 Introduction

Rapid industrialization and modernization around the world have produced the unfortunate consequence of releasing toxic wastes to the environment. Metal pollutants are derived mainly from industrial and agricultural activities. The former includes activities such as waste disposal, chemical manufacturing, and metal pollutants from vehicle exhaust, and the latter involves activities such as the use of agrochemicals, long-term application of sewage sludge, and wastewater to agricultural soils. Such releases have adversely affected human health and have produced toxic effects on plants and the soil microorganisms associated with them. Toxic metal contaminants from wastes or other products accumulate in the agricultural

---

H.I. Tak • F. Ahmad • O.O. Babalola (✉)
Department of Biological Sciences, Faculty of Agriculture, Science and Technology,
North-West University, Mafikeng Campus, Private Bag X2046, Mmabatho 2735, South Africa
e-mail: olubukola.babalola@nwu.ac.za

soils to which they are applied, threaten food security, and pose health risks to living organisms by their transfer within the food chain. Once heavy metals reach the soil, they are absorbed by plants and may be taken up by animals and humans through consumption of contaminated food or drinking water. They may even be inhaled as particulate contaminants, and due to their persistent nature, they may accumulate in both plants and animals over time.

Although plants may suffer damage from excessive contact with heavy metals, they also rely on a range of transition (heavy) metals as essential micronutrients for normal growth and development. Among metals, some comprise elements that are essential for most redox reactions and are fundamental to normal cellular functions. Fe, for example, is a key component of haem proteins such as cytochromes, catalase, and Fe-S proteins (e.g., ferredoxin). Cu is an integral component of certain electron transfer proteins in photosynthesis (e.g., plastocyanin) and respiration (e.g., cytochrome *c* oxidase) processes, whereas Mn is less redox active but has a key role in photosynthesis (e.g., O_2 evolution). Zn is non-redox active but has a vital structural and/or catalytic role in many proteins and enzymes. Ni is a constituent of urease, and small quantities of Ni are essential for some plant species (Sirko and Brodzik 2000). When amounts of any of these metals are in short supply, a range of deficiency symptoms appear and growth is reduced. Notwithstanding, these transition metals, when present in excessive amounts, interfere with cellular functions, and alter normal metabolic processes to produce cellular injuries, and potentially plant death. Many target molecules in cells, whose structure/activity is inhibited, modified or enhanced by transition metals have been identified. Ochiai (1977) lists three events that generate plant toxicity by transition (heavy) metals. These are

(a) Displacing essential components in biomolecules
(b) Blocking essential biological functions of molecules
(c) Modifying enzyme/proteins, plasma membrane, and/or membrane transporters' structure/function

Cellular enzymes and proteins contain several mercapto ligands that can structurally chelate metals, and thereby cause these proteins to lose their functional property. Heavy metals also generate oxidative stress that is mediated through generation of free radicals (Seth et al. 2008). Some heavy metals directly affect biochemical and physiological processes through inhibition of photosynthesis and respiration leading to reduced growth (Vangronsveld and Clijsters 1994). Therefore, excessive heavy-metal accumulation in plants may induce toxicity by modifying essential protein structure or replacing essential elements that is manifested by chlorosis, growth impairment, browning of roots, and inactivation of photosystems, among other effects (Shaw et al. 2004; Gorhe and Paszkowski 2006).

Governmental entities and their citizens alike have come to understand that consuming contaminated food or drinking water can expose both humans and animals to toxic levels of heavy metals. Lead poisoning is not uncommon, and is perhaps is the best example of a heavy metal the effects of which are understood by people. It has been estimated that lead exposure affects more than 800,000 children between the age of 1 and 5 in the USA (Lasat 2002). To deal with known exposures, whether from point sources, contaminated sites, or spills, efforts are routinely taken to remediate and control heavy-metal pollutants in soils and elsewhere. The three major options to remediate toxic metals are through physical, chemical, or biological means. The latter approach is referred to as bioremediation. Bioremediation is defined as the application of biological processes to remove hazardous chemicals from the environment. It has obvious advantages over physicochemical remediation in that it is more cost effective, more convenient to undertake, and produces less collateral destruction of onsite substrate or indigenous flora and fauna (Timmis and Pieper 1999).

One form of bioremediation is called phytoextraction. It is emerging as a promising and cost-effective approach to remediate metal contaminated sites, and has advantages over the alternative chemical and physical remediation technology approaches (Zhuang et al. 2007). How successful phytoremediation will be at any one site depends on the extent of soil contamination, bioavailability of the metal contaminant involved, and the ability of the plant used to absorb and accumulate metals as biomass. Generally, plants with exceptionally high metal accumulating capacity often grow slowly and produce limited biomass, particularly when the metal concentration in the soil is high. However, there is a way to maximize the chances of success of phytoremediation by utilizing plant growth-promoting rhizobacteria (PGPR), which are soil microbes that inhabit the rhizosphere. When PGPR are introduced to a contaminated site, they increase the potential for plants that grow there to sequester heavy metals and to recycle nutrients, maintain soil structure, detoxify chemicals, and control diseases and pests; PGPR also decrease the toxicity of metals by changing their bioavailability in plants. The plants, in turn, provide the microorganisms with root exudates such as free amino acids, proteins, carbohydrates, alcohols, vitamins, and hormones, which are important sources of their nutrition. The rhizosphere has high concentrations of root-exuded nutrients and attracts more bacteria than does bulk soil (Han et al. 2005; Babalola 2010). These bacteria of the rhizosphere, therefore, facilitate plant growth.

PGPR serve several functions for plants. They fix atmospheric nitrogen and supply it to plants, or they synthesize siderophores that can solubilize and sequester iron from the soil and provide it to plant cells. PGPR also synthesize several different phytohormones, including auxins (IAA) and cytokinins, which enhance plant growth and solubilize minerals such as phosphorus, thereby rendering it more readily available for plant growth. Moreover, PGPR contain enzymes that modulate plant growth and development (Glick et al. 1998; Sheng and Xia 2006; Ma et al. 2009).

The concept of using green plants to extract metals is not new. The original concept has, however, been expanded to include the use of an interdisciplinary approach, in which microbes, in close association with plants, are employed to enhance the removal, immobilization, or degradation of certain metals from polluted soils. Such enhanced action results from the presence of PGPR. PGPR comprise several different genera (e.g., *Pseudomonads* and *Acinetobacter*) and enhance the phytoremediation abilities of non-hyperaccumulating maize (*Zea mays* L.) plants by increasing their growth and biomass (Lippmann et al. 1995). Although, until recently, data on microbe-assisted metal extraction have been scarce, several papers have been published on the topic during the last 5 or 6 years. The main aim we have in this chapter is to describe how the partnership between plants and associated microbes (viz., PGPR) can be exploited as a strategy to improve plant biomass production and heavy-metal removal from metal-contaminated sites.

2 Phytoremediation

Phytoremediation is an in situ biomediation process that uses green plants and the microorganisms that are associated with them to extract, sequester, or detoxify pollutants. Plants have the capacity to take up, accumulate, degrade, or eliminate metals, pesticides, solvents, crude oil, and many industrial contaminants. Phytoremediation is a clean, cost-effective, environment-friendly technology, especially for treating large and diffused areas that are contaminated. There are many successful examples where phytoremediation has been employed, and where it has been documented to work well for remediating contaminated industrial environments (Macek et al. 2000; Suresh and Ravishankar 2004). Depending on the method used and nature of the contaminant involved, phytoremediating areas where metals and other inorganic compounds exist, may utilize one of several techniques (Glick 2003; Newman and Reynolds 2004) as follows.

(a) *Phytoextraction*: also known as phytoaccumulation, removes metals by taking advantage of the unusual ability of some plants to (hyper-)absorb and accumulate or translocate metals or/metalloids, by concentrating them within the biomass. The purpose of this type of remediation is to reduce the concentration of metals in contaminated soils so that they can be used profitably for agriculture, forestry, horticulture, grazing, etc.

(b) *Phytostabilization*: also known as phytoimmobilization, utilizes plants, often in combination with soil additives, to assist in mechanically stabilizing sites for reducing pollutant transfer to other ecosystem compartments and to the food chain; the "stabilized" organic or inorganic compound is normally incorporated into plant lignin or into soil humus. The basis for phytostabilization is that metals do not degrade, so capturing them in situ is often the best alternative. This approach is particularly applicable when low-concentration, diffused, and vast areas of contamination are to be treated. Plants restrict pollutants by creating a zone around the roots where the pollutant is precipitated and stabilized. When phytostabilization is undertaken, the plants used do not absorb the targeted pollutant(s) into plant tissue.

(c) *Phytostimulation*: plant roots promote the development of rhizosphere microorganisms that are capable of degrading the contaminant, and the microbes utilize plant root exudates as a carbon source.

(d) *Phytovolatilization/rhizovolatilization*: employs metabolic capabilities of plants and the associated rhizospheric microorganisms to transform pollutants into volatile compounds that are released into the atmosphere. Some ions (i.e., of elements of subgroups II, V, and VI of the periodic table like mercury, selenium, and arsenic) are absorbed by roots, are converted into less toxic forms, and then are released.

(e) *Phytodegradation*: organic contaminants are degraded or mineralized by specific enzymes.

(f) *Rhizofiltration*: use terrestrial plants to absorb, concentrate, and/or precipitate contaminants in the aqueous system. Rhizofiltration is also used to partially treat industrial and agricultural runoff.

Plants that can potentially accumulate large quantities of metals by natural methods have been identified, and are being studied for their use to remediate heavy-metal contaminants. These plants are called hyperaccumulators, and are often found growing in areas that have long had elevated metal concentrations in soil. Unfortunately, at high enough metal levels, even hyperaccumulating plants are slow growing and attain only a small size. Thus, high metal levels inhibit plant growth, even in plants that are capable of hyperaccumulating them. Depending upon the amount of metal at a particular site and the type of soil, even hyperaccumulating plants may require 15–20 years to remediate a contaminated site. This time frame is usually too slow for practical application. Therefore, research undertaken to find such plants should emphasize species that are fast growing and accumulate greater amounts of biomass, in addition to their being tolerant to one or more heavy metals. Moreover, the success of phytoremediation depends on the metal in the soil being in the bioavailable fraction. Hence, it is also important that researchers study the bioavailability and uptake of target metals by hyperaccumulating plants. When research is dedicated to finding optimal hyperaccumulator plants, key study goals should include both (1) evaluating the impact of metal stress on beneficial rhizospheric microbes and crops and (2) predicting the application of bioremediation technologies that could be used to clean up metals from the polluted soils.

2.1 Biological Availability of Metals in Soil

The term bioavailability is usually ill defined and rarely quantified, particularly in microbial investigations. In reality, bioavailability cannot be measured. It can only be estimated by measuring the growth of organisms, and by evaluating uptake or toxicity of the metal. As mentioned earlier, several industrial operations (e.g., smelting, mining, metal forging, manufacturing of alkaline storage batteries, and combustion of fossil fuels) release toxic metals, and agrochemicals and long-term application of sewage sludge and wastewater are used to augment agricultural production, all of which add amounts of metals to soil (Giller et al. 1989; McGrath et al. 1995; Tak et al. 2010). In soil, such metals exist in both bioavailable and non-bioavailable forms (Sposito 2000) and their mobility depends on whether they (1) precipitate in soil as positively charged ions (cations) or (2) are associated with negatively charged salt components of soil. The major reason of the low metal extracted by the plants from the soil is their low bioavailability. The bioavailability of metals from soil depends on soil factors such as cation exchange capacity (CEC), organic matter content, the content of clay minerals and hydrous metal oxides, pH, buffering capacity, redox potential, water content, and temperature (Kayser et al. 2001). In addition, plant root exudates and microbial activities in soil also influence metal bioavailability to plants (Brown et al. 1999; Traina and Laperche 1999).

Several researchers have shown the positive impact of bioaugmentation on metal bioavailability. In one such study, exchangeable Pb was increased by 113% when *Pseudomonas aeruginosa* and *Pseudomonas fluorescens* were present (Braud et al. 2006). In another study, extractable Ni was increased by the presence of *Microbacterium arabinogalactanolyticum* (Abou-Shanab et al. 2006). The toxicity of the heavy metals within soils of high CEC is generally low, even when metal concentrations are high (Roane and Pepper 2000). However, under oxidizing and aerobic conditions, metals are usually present as soluble cationic forms, whereas under reducing or anaerobic conditions they exist as sulfide or carbonate precipitates. At low soil pH, bioavailability increases due to the presence of metals as free ionic species, and just the opposite occurs when soil pH is high (i.e., when insoluble metal mineral phosphates and carbonates are formed). The soil mobility and bioavailability of the following metals usually proceed in this order: $Zn > Cu > Cd > Ni$ (Lena and Rao 1997). When metals coexist with other metals or contaminants the rate of absorption and accumulation into food webs, and ultimately into animal and human diets, may be accentuated.

2.2 Plant Uptake and Transport of Metals

Plants have developed mechanisms by which they can effectively absorb metals from the soil solution and transport them to other parts within the plant. Most metal-accumulating species were discovered in areas having a high metal concentration, and majority of such areas exist in tropical regions. These natural plant hyperaccumulators of metal represent diverse taxa, although the majority exist in the family Brassicaceae. For example, Indian mustard (*Brassica juncea*) rapidly concentrates Cd (II), Ni (II), Pb (II), and Sr (II) into root tissues at levels 500 times greater than the liquid medium in which they are growing (Salt et al. 1995; Salt and Kramer 1999). Uptake of metals into root cells, which is the point of entry into living tissue, is a major step in the phytoextraction process. However, for phytoextraction to be successful, the absorbed metals must be transported from root to shoot. The mechanisms by which metals are absorbed into the plant root are complex. This process involves transfer of metals from the soil solution to the root–surface interface, and then penetration through the root membranes to root cells. Metal ions cannot move freely across the cellular membrane because of their charge. Therefore, ion transport into cells must be mediated by membrane proteins that have a transport function, and these are generically referred to as transporters. These transporters possess an extracellular domain to which the ions attach just before the transport, and a transmembrane binding structure that connects extracellular and intracellular media.

This is an oversimplification, and the uptake process is actually rendered even more complex by the nature of the rhizosphere (Laurie and Manthey 1994).

Hyperaccumulator plants take metals up from soil in direct proportion to their bioavailability (Wenzel et al. 2003). Metal bioavailability to plants is often modified by the direct influence of root activity in the rhizosphere (Hinsinger et al. 2005). Bioavailability is regulated by electrochemical potential gradient that exists for each metal ion across the plasma membrane of root cells (Welch 1995). However, the exact nature of the membrane transporters that control the influx across the plasma membrane into the cytoplasm is not yet known.

2.3 Plant Mechanisms for Metal Detoxification

Although micronutrients such as Zn, Mn, Ni, and Cu are essential for plant growth and development, high intracellular concentrations of these ions can be toxic. To deal with this potential stress, common non-accumulator plants have evolved several mechanisms to control the intracellular homeostasis of ions (Lasat 2002; Seth et al. 2007). One such mechanism is to regulate ion influx. This involves stimulating transporter activity at a low level of intracellular ion supply, and inhibiting activity at high concentrations, or extruding ions from the cell interior to the external solution (Pollard et al. 2002). Metal hyperaccumulator species are capable of accumulating metals up to multiple thousands of ppm, and are equipped with detoxification mechanisms to address metal-induced stress. Ni(II) for example is sequestered by bonding with organic sulfur (R-SH) on the cysteine residues of peptides (Meagher 2000). Sequestration of Zn into shoot vacuoles has also been suggested to enhance tolerance in the Zn-hyperaccumulator plant *Thlaspi caerulescens* (Lasat et al. 2000). Several mechanisms have been proposed to account for Zn inactivation in vacuoles, including precipitation as Zn-phytate (Van Steveninck et al. 1990) and binding to low molecular weight organic acids (Salt 1999). Cadmium, a potentially toxic metal, also accumulates in plants, where it is detoxified by binding to phytochelatins (PC) (Vogeli-Lange and Wagner 1990; Salt and Rauser 1995), a family of thiol (SH)-rich peptides (Grill et al. 1987). Similarly, metallothionein (MT) compounds (proteins) have heavy metal-binding properties (Tomsett and Thurman 1988) and exist in numerous animals, and more recently, have been found in several plant and bacterial species (Kagi 1991). After the PC structures were elucidated, these peptides were found to be widely distributed in the plant kingdom, and were proposed to be functional equivalents of MTs (Grill et al. 1987). The role of PCs and MTs in metal detoxification has been well documented by Cobbett (2000). Apart from PCs, plants have also developed good tolerance mechanisms against the oxidative stress induced by heavy-metal exposure (Dat et al. 2000; Mishra et al. 2006).

Aerobic organisms are frequently exposed to ROS (reactive oxygen species). ROS is a generic term embracing not only free radicals, such as superoxide and hydroxyl radicals, but also H_2O_2 and singlet oxygen. These incompletely reduced oxygen species are toxic by-products, generated at low levels even in non-stressed plant cells within the chloroplasts and mitochondria during the redox reactions of photosynthetic electron transport and respiration. Stressful conditions generate more ROS; such stress may be from pathogen attacks, herbivore feeding, or exposures to UV light, heavy metals, or other substances (Diaz et al. 2001). Heavy-metal induced stress is often associated with degenerative reactions that are primarily associated with ROS; it is well known that their production in plant cells must be minimized to limit the effects of their high reactivity (Noctor and Foyer 1998). Toxicity generated by both O_2 and H_2O_2 is presumed to result from their ability to initiate reaction cascades that produce hydroxyl radicals capable of causing lipid peroxidation, protein denaturation, and DNA mutations. ROS may also damage the pigment system and the chlorophyll (Schutzendubel and Polle 2002; Sudo et al. 2008). Notwithstanding, nature has equipped plants with active antioxidant systems, which scavenge the toxicity generated by ROS (Cobbett 2000; Hou et al. 2007; Skorzynska-Polit et al. 2010). Such scavenging occurs by the synchronous action of several antioxidant enzymes that includes superoxidase dismutase (SOD), which converts superoxide to H_2O_2 (Bowler et al. 1992). Catalases (CATs) help convert H_2O_2 to water and molecular oxygen in peroxisomes (Noctor and Foyer 1998). In addition, an alternative mode of H_2O_2 destruction exists via the action of peroxidases (POD), which are found throughout the cell and have a much greater affinity for H_2O_2 than do CATs (Jimenez et al. 1997). Enzymes that exist in the ascorbate–glutathione cycle, where H_2O_2 is scavenged, are also highly active. In this cycle, the ascorbate peroxidases (APX) catalyze the reduction of H_2O_2 to water by ascorbate, and the resulting dehydroascorbate is reduced back to ascorbate with help from glutathione reductases (GRs) (Iturbe-Ormaetxe et al. 2001). Although plants have the ability to combat negative consequences of heavy-metal stress via the presence of antioxidants, many studies have shown that exposure to elevated concentrations of reactive metals reduce, rather than increase, antioxidative enzyme activities (Schutzendubel and Polle 2002). Because multiple factors affect heavy-metal tolerance in plants it is difficult to elucidate the exact tolerance mechanisms involved. However, two promising strategies are emerging to remediate polluted sites; the first is plant growth enhancement, and the second is reduced metal translocation, both of which are caused by soil amendments and/or microbial inoculations (PGPR) to multicontaminated sites.

3 Plant Growth-Promoting Bacteria

The beneficial free-living soil bacteria that exist in association with the roots of many different plants are generally referred to as plant growth-promoting rhizobacteria (PGPR) (Kloepper and Schroth 1978). Depending on their relationship with the host plants, PGPR can be divided into two major groups: (1) symbiotic rhizobacteria, which may invade the interior of cells and survive inside the cell (also called intracellular PGPR, e.g., nodule bacteria), and (2) free-living rhizobacteria that exist outside plant cells (called extracellular PGPR, e.g., *Bacillus*, *Pseudomonas*, *Burkholderia*, and *Azotobacter*) (Khan 2005; Babalola and Akindolire 2011). The major factor that affects the high concentration of bacteria found in the rhizosphere

Table 1 Plant growth-promoting rhizobacteria and the growth-regulating compounds associated with them

Organisms	Growth regulators	References
Pseudomonas chlororaphis, Arthrobacter pascens	Phosphate solubilization	Yu et al. (2012)
Pseudomonas spp.	Indole acetic acid (IAA), siderophore, P-solubilization	Li and Ramakrishna (2011)
Achromobacter xylosooxidans	IAA, P solubilisation	Ma et al. (2009)
Pseudomonas spp., *Bacillus megaterium*	IAA, siderophore and P solubilization	Rajkumar and Freitas (2008)
Microbacterium G16	IAA, siderophores	Sheng et al. (2008)
Azotobacter, Fluorescent pseudomonas, and *Bacillus*	IAA, siderophore, ammonia, HCN, P-solubilization	Ahmad et al. (2008)
Bacillus spp.	IAA, P solubilization, siderophores, HCN, ammonia	Wani et al. (2007)
Pseudomonas and *Bacillus*	Siderophores, IAA, P-solubilization	Rajkumar et al. (2006)
Brevibacillus brevis	IAA	Vivas et al. (2006)
Bravibacterium sp.	Siderophore	Noordman et al. (2006)
Xanthomonas sp. *RJ3, Azomonas* sp. *RJ4, Pseudomonas* sp. *RJ10, Bacillus* sp. *RJ31*	IAA	Sheng and Xia (2006)
Bacillus subtilis	IAA and P-solubilization	Zaidi et al. (2006)
Bacillus sp.	P-solubilization	Canbolat et al. (2006)
Variovorax paradoxus, Rhodococcus sp. and *Flavobacterium* (Cd tolerant)	IAA and siderophores	Belimov et al. (2005)
Sphingomonas sp., *Mycobacterium* sp., *Bacillus* sp., *Rhodococcus* sp., *Cellulomonas* sp. and *Pseudomonas* sp.	IAA	Tsavkelova et al. (2005)
Micrococcus luteus	IAA, P-solubilization	Antoun et al. (2004)
Bacillus, Pseudomons, Azotobacter, and *Azospirillum*	P-solubilization and IAA	Tank and Saraf (2003)
Pseudomonas fluorescence	Siderophore	Khan et al. (2002)
Kluyvera ascorbata	Siderophore	Burd et al. (2000)

is the presence of high nutrient levels (especially small molecules such as amino acids, sugars, and organic acids) that are exuded from the roots of most plants (Bayliss et al. 1997; Penrose and Glick 2001). PGPR can positively influence plant growth and development in three different ways, in that they

1. Synthesize and provide growth-promoting compounds to the plants (Glick 1995) (Table 1),
2. Facilitate the uptake of certain environmental nutrients such as nitrogen, phosphorus, sulfur, magnesium, and calcium (Bashan and Levanony 1990; Belimov and Dietz 2000; Cakmakci et al. 2006), and
3. Decrease or prevent some deleterious effects caused by phytopathogenic organisms or other diseases (Khan et al. 2002; Lugtenberg and Kamilova 2009)

Generally, rhizobacteria improve plant growth by synthesizing phytohormone precursors (Ahmad et al. 2008), vitamins, enzymes, siderophores, and antibiotics (Burd et al. 2000; Noordman et al. 2006). PGPR also increase plant growth by synthesizing specific enzymes, which induce biochemical changes in plants. For example, ethylene plays a critical role in various plant developmental processes, such as leaf senescence and abscission, epinasty, and fruit ripening (Vogel et al. 1998). Ethylene also regulates node factor signaling, nodule formation, and has primary functions in plant defense systems. Moreover, as a result of the plant infection by rhizobacteria, ethylene production is increased (Boller 1991), which, at higher concentrations, will inhibit plant growth and development (Morgan and Drew 1997; Grichko and Glick 2001). However, bacterial 1-aminocyclopropane-1-carboxylate (ACC), a deaminase synthesized by PGPR (Babalola et al. 2003; Madhaiyan et al. 2006; Rajkumar et al. 2006), alleviates stress induced by such ethylene-mediated impact. In addition, rhizobacterial strains can solubilize inorganic P (Glick et al. 1998; Yu et al. 2012), or mineralize organic P, and thereby improve plant stress tolerance to drought, salinity, and metal toxicity (Ponmurugan 2006; Khan et al. 2007). Pishchik et al. (2005) mathematically simulated the succession of events that began with phytohormone (IAA and ethylene) synthesis and ended with higher uptake of ions by roots, under conditions of cadmium stress. Possibly, synthesis of phytohormones might be stimulated by exposure to heavy metals. Conversely, these processes may be hindered by high heavy-metal concentrations (DellAmico et al. 2005), because many rhizobacteria cannot survive when such concentrations are high. These authors further reported that many different microbial communities are able to withstand high heavy-metal concentrations when living in association with rhizospheric soils and the rhizoplane.

The mechanism of metal tolerance and the possible metal transforming capacities of the metal-resistant PGPR are briefly discussed in the following section.

3.1 How Do PGPR Combat Heavy-Metal Stress?

Unlike many pollutants, which undergo biodegradation to produce less toxic, less mobile, and less bioavailable products, removing heavy metals from a contaminated environment is much more difficult. Heavy metals cannot be degraded biologically and are ultimately indestructible, though the speciation and bioavailability of metals may change as environmental factors change. Some metals (e.g., zinc, copper, nickel, and chromium) are essential and beneficial micronutrients for plants, animals, and microorganisms (Olson et al. 2001), whereas others (e.g., cadmium, mercury, and lead) have no known biological or physiological function (Gadd 1992). However, high concentrations of these heavy metals greatly affect microbial communities, and may reduce their total microbial biomass (Giller et al. 1998), their activity (Romkens et al. 2002), or change microbial community structure (Gray and Smith 2005). Therefore, at higher concentrations, either heavy-metal ions completely inhibit a microbial population by inhibiting its various metabolic activities or these organisms may develop resistance or tolerance to such elevated levels of heavy metals. This ability to live and grow under in the presence of high metal concentration exists in many rhizospheric microorganisms. Ledin (2000) explained the difference between microbial tolerance and resistance; he defines tolerance as the ability to cope with metal toxicity by means of intrinsic properties of the microorganisms, whereas resistance is the ability of microbes to detoxify heavy metals by being activated in direct response to the high heavy-metal concentrations.

Toxic heavy-metal pollutants should be either completely removed from the contaminated soil or transformed or immobilized in ways that render them safe. For survival under metal-stressed environment, PGPR have developed a range of mechanisms by which they can immobilize, mobilize, or transform heavy metals, thereby rendering them inactive (Nies 1999). These mechanisms include (1) exclusion-metal ions that are kept away from target sites, (2) extrusion-metals that are pushed out of the cell through chromosomal/plasmid mediated events (3) accommodation—metals that form complexes with metal-binding proteins, e.g., mettalothioneins, low molecular weight proteins (Kao et al. 2006; Umrania 2006), and other cell components, (4) biotransformation, in which the toxic metal is reduced to less toxic forms, and (5) methylation and demethylation. One or more of the above-mentioned mechanisms allow the microbes to function metabolically in metal-contaminated sites/soils. Interest in exploiting these bacterial properties to remediate heavy-metal contaminated sites is growing, and early results from their application are promising (Lloyd and Lovley 2001; Hallberg and Johnson 2005).

3.2 Synergistic Interaction of PGPR and Plants in Heavy-Metal Remediation

Although many plant–microbe interactions have been investigated, the studies performed so far have mainly emphasized plant–pathogen interactions. Only 10 years ago, research on the ecology of microbes in the rhizosphere was focused on the microbiological detoxification and decontamination of soil as affected by heavy metals. The fact that PGPR promotes plant growth is well documented (Reed and Glick 2004; Babalola et al. 2007; Babalola 2010), and more recently, PGPR have been successfully used to reduce plant stress in metal-contaminated soils. The microorganisms that are associated with roots establish a synergistic relationship with plant roots which enhances nutrient absorption and improves plant performance, as well as the quality of soils (Tinker 1984; Yang et al. 2009). Bacteria interact with and affect plant growth in a variety of ways. Some bacteria are phytopathogenic and actively inhibit plant growth; others (e.g., PGPR) facilitate plant growth through several mechanisms; many soil bacteria do not appear to affect the plant growth at all, although a change in soil conditions could reverse this (Glick 1995). Some microbial communities have the ability to sequester heavy metals, and therefore may be useful for bioremediating contaminated areas (Hallberg and Johnson 2005; Umrania 2006). When microbes are used to bioremediate a contaminated site, plant-associated bacteria can be potentially used to improve phytoextraction activities by altering the solubility, availability, and transport of heavy metals, and nutrients as well, by reducing soil pH and releasing chelators (Ma et al. 2011). Among the metabolites produced by PGPR, siderophores play a significant role in metal mobilization and accumulation (Rajkumar et al. 2010). Recently, Cr and Pb were found to be released into the soil solution after soil was inoculated with *P. aeruginosa* (Braud et al. 2009). *P. aeruginosa* can realistically only serve as a model system, because it is a well-known pathogen, and regulators would not allow deliberate release of it to the environment. Although no field success has yet been achieved by doing so, the concept of inoculating seeds/rhizospheric soils with selected metal-mobilizing bacteria to improve phytoextraction in metal-contaminated soils has merit.

3.3 ACC Deaminase and Plant Stress Reduction from Ethylene

As mentioned earlier, ethylene is produced under normal plant-growth conditions and regulates plant growth, although it is toxic to plants at higher concentrations (Bestwick and Ferro 1998). Ethylene is produced from L-methionine via the intermediates *S*-adenosyl-L-methionine (SAM) and 1-aminocyclopropane-1-carboxylic acid (ACC) (Yang and Hoffman 1984). In a step-wise metabolic reaction, methionine is first converted into *S*-adenosyl-L-methionine by SAM synthetase (Giovanelli et al. 1980) and SAM is then hydrolyzed to ACC and 5-methyl thioadenosine (Kende 1989) by ACC synthetase. Finally, ACC is metabolized to ethylene, CO_2, and cyanide by ACC oxidase (John 1991). Several chemicals have been used to control ethylene-mediated stress in plants. Unfortunately, use of such substances is environmentally unfriendly. Applying cyclopropenes can block the action of ethylene and it can potentially be used to extend the shelf life of flowers, and potted plants. Other compounds are known to inhibit ethylene biosynthesis, although they are potentially harmful to the environment, e.g., silver thiosulfate (Bestwick and Ferro 1998).

Generally, it is important that ethylene levels in plants be kept at minimum levels possible. This can be accomplished by minimizing the ethylene precursor ACC, which is subject to degradation by an ACC enzyme isolated from a *Pseudomonas* sp. strain ACP and from yeast *Hansenula saturnus* (Honma and Shimomura 1978; Minami et al. 1998). An ACC deaminase has also been detected in the fungus *Penicillium citrinum*, and bacterial strains originating from the soil that have ACC deaminase activity have been reported (Glick 1995; Jia et al. 2000; Belimov et al. 2001; Babalola et al. 2003). This enzyme, ACC deaminase, degrades ACC to ammonia and α-ketobutyrate, and can be utilized to protect plants from ethylene-generated stress (Glick et al. 1998). Many plant species require ethylene for seed germination, and ethylene production increases during seed germination and seedling growth (Abeles et al. 1992). However, elevated ethylene levels may inhibit root elongation and depress growth (Morgan and Drew 1997). In higher plants,

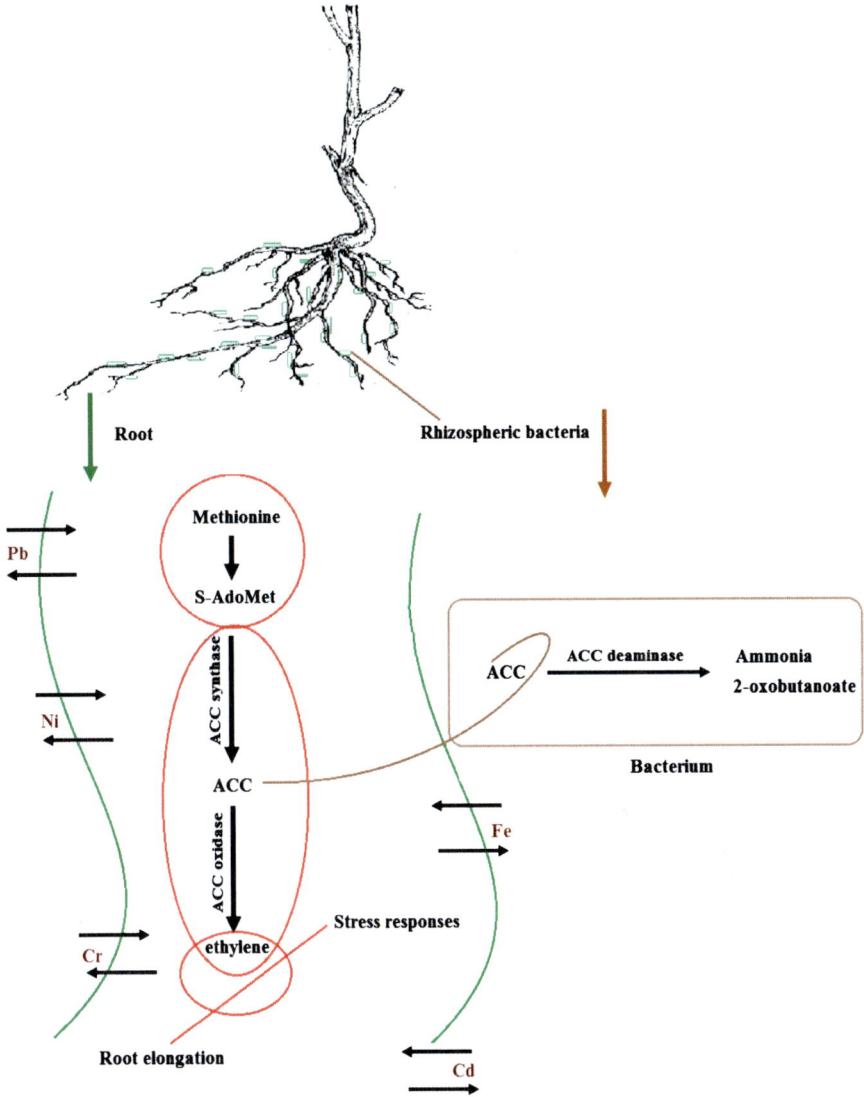

Fig. 1 Diagrammatic model showing the process for reducing ethylene levels in roots by using bacterial 1-aminocyclopropane-1-carboxylic acid (ACC) deaminase. ACC synthesized in plant tissues is believed to be exuded from plant roots and is taken up by rhizobacteria where ACC is hydrolyzed to ammonia and 2-oxobutanoate

S-Adenosyl-L-methionine (S-AdoMet) is synthesized from methionine and ACC is synthesized and converted into ethylene by ACC oxidase. Further ethylene production in plants is controlled by regulating the expression of ACC synthase and ACC oxidase genes (Kim et al. 2001). In Fig. 1, a model of the process by which ethylene levels are reduced in roots is presented, along with the role played by the ACC enzyme (Glick et al. 2007).

PGPR synthesize the indole-3-acetic acid (IAA) utilizing tryptophan excreted by roots in the rhizospheric region. The synthesized IAA molecules are then secreted and transported into plant cells. These auxins have dual roles. One is to participate in plant cell growth and the other is to promote ACC synthase activity to increase the ethylene titer. Stress induces an increase in ACC levels and, therefore, emulates the action of IAA molecules. Increased ACC molecules then diffuse from plants and are imported into PGPR cells where they are subjected to the action of ACC deaminase. Because of this, microbes and plants are more tolerant to stress-induced growth inhibition that is mediated by ethylene. When tested, strains of ACC deaminase-containing plant growth-promoting bacteria were found to reduce the amount of ACC that was detectable by HPLC, and the ethylene levels in canola seedlings were also lowered (Penrose and Glick 2001). Thus PGPR can potentially be used to counter ethylene-mediated stress, although field trials are needed to elucidate the mechanism by which this occurs.

4 Summary

In this review, we briefly describe the biological application of PGPR for purposes of phytoremediating heavy metals. We address the agronomic practices that can be used to maximize the remediation potential of plants. Plant roots have limited ability to absorb metals from soil, mainly because metals have low solubility in the soil solution. The phytoavailability of metal is closely tied to the soil properties and the metabolites that are released by PGPR (e.g., siderophores, organic acids, and plant growth regulators). The role played by PGPR may be accomplished by their direct effect on plant growth dynamics, or indirectly by acidification, chelation, precipitation, or immobilization of heavy metals in the rhizosphere.

From performing this review we have formed the following conclusions:

- The most critical factor in determining how efficient phytoremediation of metal-contaminated soil will be is the rate of uptake of the metal by plants. In turn, this depends on the rate of bioavailability. We know from our review that beneficial bacteria exist that can alter metal bioavailability of plants. Using these beneficial bacteria improves the performance of phytoremediation of the metal-contaminated sites.
- Contaminated sites are often nutrient poor. Such soils can be nutrient enriched by applying metal-tolerant microbes that provide key needed plant nutrients. Applying metal-tolerant microbes therefore may be vital in enhancing the detoxification of heavy-metal-contaminated soils (Glick 2003).
- Plant stress generated by metal-contaminated soils can be countered by enhancing plant defense responses. Responses can be enhanced by alleviating the stress-mediated impact on plants by enzymatic hydrolysis of ACC, which is intermediate in the biosynthetic pathway of ethylene. These plant–microbe partnerships can act as decontaminators by improving phytoremediation.

Soil microorganisms play a central role in maintaining soil structure, fertility, and in remediating contaminated soils. Although not yet widely applied, utilizing a plant–microbe partnership is now being recognized as an important tool to enhance successful phytoremediaton of metal-contaminated sites. Hence, soil microbes are essential to soil health and sustainability. The key to their usefulness is their close association with, and positive influence on, plant growth and function. To capitalize on the early success of this technique and to improve it, additional research is needed on successful colonization and survival of inoculums under field conditions, because these are vital for the success of this approach. In addition, the effects of the interaction of PGPR and plant root-mediated process on the metal mobilization in soil are required, to better elucidate the mechanism that underlies bacterial-assisted phytoremediation. Finally, applying PGPR-associated phytoremediation under field conditions is important, because, to date, only locally contaminated sites have been treated with this technique, by using microbes cultured in the laboratory.

Acknowledgements The first author is thankful to the North-West University for the award of postdoctoral fellowship. This work is based on research supported by the National Research Foundation.

References

Abeles FB, Morgan PW, Sltveit ME Jr (1992) Ethylene in plant biology. Academic, New York, p 414

Abou-Shanab RAI, Angle JS, Chaney RL (2006) Bacterial inoculants affecting nickel uptake by Alyssum murale from low, moderate and high Ni soils. Soil Biol Biochem 38:2882–2889

Ahmad F, Ahmad I, Khan MS (2008) Screening of free-living rhizospheric bacteria for their multiple plant growth promoting activities. Microbiol Res 163:173–181

Antoun H, Beauchamp CJ, Goussard N, Chabot R, Llande R (2004) Potential of *Rhizobium* and *Bradyrhizobioum* species as plant growth promoting rhizobacteria on non-legumes: effect on radishes. Plant Soil 204:57–67

Babalola OO (2010) Beneficial bacteria of agricultural importance. Biotechnol Lett 32(11): 1559–1570

Babalola OO, Akindolire AM (2011) Identification of native rhizobacteria peculiar to selected food crops in Mmabatho municipality of South Africa. Biol Agric Hort 27(3–4):294–309

Babalola OO, Osir EO, Sanni AI, Odhiambo GD, Bulimo WD (2003) Amplification of 1-aminocyclopropane-1-carboxylic (ACC) deaminase from plant growth promoting bacteria in Striga-infested soil. Afr J Biotechnol 2:157–160

Babalola OO, Berner DK, Amusa NA (2007) Evaluation of some bacterial isolates as germination stimulants of *Striga hermonthica*. Afr J Agric Res 2(1):27–30

Bashan Y, Levanony H (1990) Current status of *Azospirillum* inoculation technology: Azospirillum as a challenge for agriculture. Can J Microbiol 36:591–608

Bayliss C, Bent E, Culham DE, MacLellan S, Clarke AJ, Brown GL, Wood JM (1997) Bacterial genetic loci implicated in the *Pseudomonas putida* GR12-2R3-canola mutualism: identification of an exudate-inducible sugar transporter. Can J Microbiol 43:809–818

Belimov AA, Dietz KJ (2000) Effect of associative bacteria on element composition of barley seedlings grown in solution culture at toxic cadmium concentrations. Microbiol Res 155:113–121

Belimov AA, Safronova VI, Sergeyeva TA, Egorova TN, Matveyeva VA, Tsyganov VE, Borisov AY, Tikhonovich IA, Kluge C, Preisfeld A, Dietz KJ, Stepanok VV (2001) Characterisation of plant growth-promoting rhizobacteria isolated from polluted soils and containing 1-aminocyclopropane-1-carboxylate deaminase. Can J Microbiol 47:642–652

Belimov AA, Hontzeas N, Safronova VI, Demchinskaya SV, Piluzza G, Bullitta S, Glick BR (2005) Cadmium-tolerant plant growth promoting rhizobacteria associated with the roots of Indian mustard (*Brassica juncea* L. Czern.). Soil Biol Biochem 37:241–250

Bestwick RK, Ferro AJ (1998) Reduced ethylene synthesis and delayed fruit ripening in transgenic tomatoes expressing S-adenosylmethionine hydrolase. US Patent 5, 723,746

Boller T (1991) Ethylene in pathogenesis and disease resistance. In: Suttle JC, Mattoo AK (eds) The plant hormone ethylene. CRC Press, Boca Raton, FL, pp 293–314

Bowler C, Van Montagu M, Inzé D (1992) Superoxide dismutase and stress tolerance. Annu Rev Plant Physiol Plant Mol Biol 43:83–116

Braud A, Jezequel K, Vieille E, Tritter A, Lebeau T (2006) Changes in extractability of Cr and Pb in a polycontaminated soil after bioaugmentation with microbial producers of biosurfactants, organic acids and siderophores. Water Air Soil Pollut: Focus 6:261–279

Braud A, Jezequel K, Bazot S, Lebeau T (2009) Enhanced phytoextraction of an agricultural Cr-, Hg and Pb-contaminated soil by bioaugmentation with siderophore producing bacteria. Chemosphere 74:280–286

Brown GE Jr, Foster AL, Ostergren JD (1999) Mineral surfaces and bioavailability of heavy metals: a molecular-scale perspective. Proc Natl Acad Sci U S A 96:3388–3395

Burd GI, Dixon DG, Glick BR (2000) Plant growth promoting bacteria that decrease heavy metal toxicity in plants. Can J Microbiol 46:237–245

Cakmakci R, Donmez F, Aydm A, Sahin F (2006) Growth promotion of plants by plant growth promoting rhizobacteria under greenhouse and two different field soil conditions. Soil Biol Biochem 38:1482–1487

Canbolat MY, Bilen S, Cakmakci R, Sahin F, Aydin A (2006) Effect of plant growth promoting bacteria and soil compaction on barley seedling growth, nutrient uptake, soil properties and rhizosphere microflora. Biol Fertil Soils 42:350–357

Cobbett CS (2000) Phytochelatins and their roles in heavy metal detoxification. Plant Physiol 123:825–832

Dat JF, Van Breusegem F, Vandenabeele S, Vranova E, Van Montague M, Inze D (2000) Dual action of active oxygen species during plant stress responses. Cell Mol Life Sci 57:779–795

DellAmico E, Cavalca L, Andreoni V (2005) Analysis of rhizobacterial communities in perennial Graminaceae from polluted water meadow soil, and screening of metal-resistant, potentially plant growth-promoting bacteria. FEMS Microbiol Ecol 52:153–162

Diaz J, Bernal A, Pomar F, Merino F (2001) Induction of skhikimate dehydrogenase and peroxidase in pepper (*Capsicum annuum* L.) seedlings in response to copper stress and its relation to lignification. Plant Sci 161:179–188

Gadd GM (1992) Metals and microorganisms: a problem of definition. FEMS Microbiol Lett 100:197–204

Giller KE, McGrath SP, Hirsch PR (1989) Absence of nitrogen fixation in clover grown on soil subject to long-term contamination with heavy metals is due to survival of only ineffective *Rhizobium*. Soil Biol Biochem 21:841–848

Giller KE, Witter E, McGrath SP (1998) Toxicity of heavy metals to microorganisms and microbial process in agricultural soils: a review. Soil Biol Biochem 30:1389–1414

Giovanelli J, Mudd SH, Dakto AH (1980) Sulphur amino acids in plants. In: Miflin BJ (ed) Amino acids and derivatives, the biochemistry of plants: a comprehensive treatise, vol 5. Academic, New York, pp 453–505

Glick BR (1995) The enhancement of plant growth by free-living bacteria. Can J Microbiol 41:109–117

Glick BR (2003) Phytoremediation: synergistic use of plants and bacteria to clean up the environment. Biotechnol Adv 21:383–393

Glick BR, Penrose DM, Li J (1998) A model for the lowering of plant ethylene concentrations by plant growth promoting bacteria. J Theor Biol 190:63–68

Glick BR, Cheng Z, Czarny J, Duan J (2007) Promotion of plant growth by ACC deaminase-containing soil bacteria. Eur J Plant Pathol 119:329–339

Gorhe V, Paszkowski U (2006) Contribution of arbuscular mycorrhizal symbiosis to heavy metal phytoremediation. Planta 223:1115–1122

Gray EJ, Smith DL (2005) Intracellular and extracellular PGPR: commonalities and distinctions in the plant-bacterium signaling processes. Soil Biol Biochem 37:395–412

Grichko VP, Glick BR (2001) Amelioration of flooding stress by ACC deaminase containing plant growth promoting rhizobacteria. Plant Physiol Biochem 39:11–17

Grill E, Winnacker E-L, Zenk MH (1987) Phytochelatins, a class of heavy-metal-binding peptides from plants are functionally analogous to metallothioneins. Proc Natl Acad Sci U S A 84:439–443

Hallberg KB, Johnson DB (2005) Microbiology of a wetland ecosystem constructed to remediate mine drainage from a heavy metal mine. Sci Total Environ 338:53–66

Han J, Sun L, Dong X, Cai Z, Sun X, Yang H, Wang Y, Song W (2005) Characterization of a novel plant growth-promoting bacteria strain *Delftia tsuruhatensis* HR4 both as a diazotroph and a potential biocontrol agent against various plant pathogens. Syst Appl Microbiol 28:66–76

Hinsinger P, Gobran GR, Gregory PJ, Wenzel WW (2005) Rhizosphere geometry and heterogeneity arising from root-mediated physical and chemical processes. New Phytol 168:293–303

Honma M, Shimomura T (1978) Metabolism of 1-amino-cyclopropane-1-carboxylic acid. Agric Biol Chem 42:1825–1831

Hou W, Chen X, Song G, Wang Q, Chang CC (2007) Effects of copper and cadmium on heavy metal polluted water body restoration by duckweed (*Lemna minor*). Plant Physiol 107:1059–1066

Iturbe-Ormaetxe I, Matamoros MA, Rubio MC, Dalton DA, Becana M (2001) The antioxidants of legume nodule mitochondria. Mol Plant Microbe Interact 14:1189–1196

Jia YJ, Ito H, Matsui H, Honma M (2000) 1-aminocyclop- ropane-1-carboxylate (ACC) deaminase induced by ACC synthesized and accumulated in *Penicillium citrinum* intracellular spaces. Biosci Biotechnol Biochem 64:299–305

Jimenez A, Hernandez JA, del Rio LA, Sevilla F (1997) Evidence for the presence of the ascorbate-glutathione cycle in mitochondria and peroxisomes of pea leaves. Plant Physiol 114:275–284

John P (1991) How plant molecular biologists revealed a surprising relationship between two enzymes, which took an enzyme out of a membrane where it was not located, and put it into the soluble phase where it could be studied. Plant Mol Biol Rep 9:192–194

Kagi JHR (1991) Overview of metallothionein. Methods Enzymol 205:613–626

Kao PH, Huang CC, Hseu ZY (2006) Response of microbial activities to heavy metals in a neutral loamy soil treated with biosolid. Chemosphere 64:63–70

Kayser G, Korckritz T, Markert B (2001) Bioleaching for the decontamination of heavy metals polluted soils with *Thiobacillus* spp. Wasser Boden 53:54–58

Kende H (1989) Enzymes of ethylene biosynthesis. Plant Physiol 91:1–4

Khan AG (2005) Role of soil microbes in the rhizospheres of plants growing on trace metal contaminated soils in phytoremediation. J Trace Elem Med Biol 18:355–364

Khan MS, Zaidi A, Aamil M (2002) Biocontrol of fungal pathogens by the use of plant growth promoting rhizobacteria and nitrogen fixing microorganisms. Indian J Bot Soc 81:255–263

Khan MS, Zaidi A, Wani PA (2007) Role of phosphate solubilizing microorganisms in sustainable agriculture:a review. Agron Sustain Dev 27:29–43

Kim JH, Kim WT, Kang BG (2001) IAA and N6-benzyladenine inhibit ethylene-regulated expression of ACC oxidase and synthase genes in mungbean hypocotyls. Plant Cell Physiol 42:1056–1061

Kloepper JW, Schroth MN (1978) Plant growth promoting rhizobacteria on radishes, fourth international conference on plant pathogen bacteria, vol 2. Angers, France, pp 879–882

Lasat MM (2002) Phytoextraction of toxic metals: a review of biological mechanisms. J Environ Qual 31:109–120

Lasat MM, Pence NS, Garvin DF, Ebbs SD, Kochian LV (2000) Molecular physiology of zinc transport in the Zn hyperaccumulator *Thlaspi caerulescens*. J Exp Bot 51:71–79

Laurie SH, Manthey JA (1994) The chemistry and role of metal ion chelation in plant uptake processes. In: Manthey JA, Crowley DE, Luster DG (eds) Biochemistry of metal micronutrients in the rhizosphere. Lewis Publishers, Boca Raton, pp 165–182

Ledin M (2000) Accumulation of metals by microorganisms-processes and importance for soil system. Earth Sci Rev 51:1–31

Lena QM, Rao GN (1997) Heavy metals in the environment. J Environ Qual 26:264

Li K, Ramakrishna W (2011) Effect of multiple metal resistant bacteria from contaminated lake sediments on metal accumulation and plant growth. J Hazard Mater 189:531–539

Lippmann B, Leinhos V, Bergmann H (1995) Influence of auxin producing rhizobacteria on root morphology and nutrient accumulation of crops. 1. Changes in root morphology and nutrient accumulation in maize (*Zea-mays* L.) caused by inoculation with indole-3-acetic acid (IAA) producing *Pseudomonas* and *Acinetobacter* strains or IAA applied exogenously. Angew Bot 69:31–36

Lloyd JR, Lovley DR (2001) Microbial detoxification of metals and radionuclides. Curr Opin Biotechnol 12:248–253

Lugtenberg B, Kamilova F (2009) Plant-growth-promoting rhizobacteria. Annu Rev Microbiol 63:541–556

Ma Y, Rajkumar M, Freitas H (2009) Inoculation of plant growth promoting bacteria *Achromobacter xylosoxidans* strain Ax10 for improvement of copper phytoextraction by *Brassica juncea*. J Environ Manage 90:831–837

Ma Y, Prasad MNV, Rajkumar M, Freitas H (2011) Plant growth promoting rhizobacteria and endophytes accelerate phytoremediation of metalliferous soils. Biotechnol Adv 29:248–258

Macek T, Macková M, Kas J (2000) Exploitation of plants for the removal of organics in environmental remediation. Biotechnol Adv 18:23–34

Madhaiyan M, Poonguzhali S, Ryu JH, Sa TM (2006) Regulation of ethylene levels in canola (*Brassica campestris*) by 1-aminocyclopropane- 1-carboxylate deaminase-containing *Methylobacterium fujisawaense*. Planta 224:268–278

McGrath SP, Chaudri AM, Giller KE (1995) Long-term effects of metals in sewage sludge on soils, microorganisms and plants. J Ind Microbiol 14:94–104

Meagher RB (2000) Phytoremediation of toxic elemental and organic pollutants. Curr Opin Plant Biol 3:153–162

Minami R, Uchiyama K, Murakami T, Kawai J, Mikami K, Yamada T, Yokoi D, Ito H, Matsui H, Honma M (1998) Properties, sequence, and synthesis in *Escherichia coli* of 1-aminocyclopropane-1-carboxylate deaminase from *Hansenula saturnus*. J Biochem 123:1112–1118

Mishra S, Srivastava S, Tripathi RD, Kumar R, Seth CS, Gupta K (2006) Lead detoxification by coontail (*Ceratophyllum demersum* L.) involves induction of phytochelatins and antioxidant system in response to its accumulation. Chemosphere 65:1027–1039

Morgan PW, Drew CD (1997) Ethylene and plant responses to stress. Physiol Plant 100:620–630

Newman LA, Reynolds CM (2004) Phytodegradation of organic compounds. Curr Opin Biotechnol 15:225–230

Nies DH (1999) Microbial heavy metal resistance. Appl Microbiol Biotechnol 51:730–750

Noctor G, Foyer CH (1998) Ascorbate and glutathione: keeping active oxygen under control. Annu Rev Plant Physiol Plant Mol Biol 49:249–279

Noordman WH, Reissbrodt R, Bongers RS, Rademaker ILW, Bockelmann W, Smit G (2006) Growth stimulation of *Brevibacterium* sp. by siderophores. J Appl Microbiol 101:637–646

Ochiai EI (1977) Bioinorganic chemistry: an introduction. Allyn and Bacon, Boston, pp 218–262

Olson JW, Mehta NS, Maier RJ (2001) Requirement of nickel metabolism protein HypA and HypB for full activity of both hydrogenase and urease in *Helicobacter pylori*. Mol Microbiol 39:176–182

Penrose DM, Glick BR (2001) Levels of 1-aminocyclopropane-1-carboxylic acid (ACC) in exudates and extracts of canola seeds treated with plant growth-promoting bacteria. Can J Microbiol 47:368–372

Pishchik VN, Vorobev NI, Provorov NA (2005) Experimental and mathematical simulation of population dynamics of rhizospheric bacteria under conditions of cadmium stress. Microbiology 74:735–740

Pollard JA, Powell KD, Harper FA, Smith JAC (2002) The genetic basis of metal hyperaccumulation in plants. Crit Rev Plant Sci 21:539–566

Ponmurugan PGC (2006) In vitro production of growth regulators and phosphatase activity by phosphate solubilizing bacteria. Afr J Biotechnol 5:340–350

Rajkumar M, Freitas H (2008) Effects of inoculation of plant-growth promoting bacteria on Ni uptake by Indian mustard. Biores Technol 99:3491–3498

Rajkumar M, Nagendran R, Lee KJ, Lee WH, Kim SZ (2006) Influence of plant growth promoting bacteria and Cr^{6+} on the growth of Indian mustard. Chemosphere 62:741–748

Rajkumar M, Ae N, Prasad MNV, Freitas H (2010) Potential of siderophore-producing bacteria for improving heavy metal phytoextraction. Trends Biotechnol 28:142–149

Reed MLE, Glick BR (2004) Applications of free living plant growth promoting rhizobacteria. Antonie Van Leeuwenhoek 86:1–25

Roane TM, Pepper IL (2000) Microorganisms and metal pollution. In: Maier RM, Pepper IL, Gerba CB (eds) Environmental microbiology. Academic, London, p 55

Romkens P, Bouwman L, Japenga J, Draaisma C (2002) Potentials and drawbacks of chelate-enhanced phytoremediation of soils. Environ Pollut 116:109–121

Salt DE (1999) Zinc ligands in the metal hyperaccumulator *Thlaspi caerulescens* as determined using X-ray absorption spectroscopy. Environ Sci Technol 33:713–717

Salt DE, Kramer U (1999) Mechanisms of metal hyperaccumulation in plants. In: Raskin I, Enslely BD (eds) Phytoremediaton of toxic metals: using plants to clean-up the environment. John Wiley and Sons, New York, pp 231–246

Salt DE, Rauser WE (1995) Mg ATP-dependent transport of phytochelatins across the tonoplast of oat roots. Plant Physiol 107:1293–1301

Salt DE, Prince RC, Pickering IJ, Raskin I (1995) Mechanisms of cadmium mobility and accumulation in Indian mustard. Plant Physiol 109:1427–1433

Schutzendubel A, Polle A (2002) Plant responses to abiotic stresses: heavy-metal induced oxidative stress and protection by mycorrhization. J Exp Bot 53:1351–1365

Seth CS, Chaturvedi PK, Misra V (2007) Toxic effect of arsenate and cadmium alone and in combination on giant duckweed (*Spirodela polyrrhiza* L.) in response to its accumulation. Environ Toxicol 22:539–549

Seth CS, Chaturvedi PK, Misra V (2008) The role of phytochelatins and antioxidants in tolerance to Cd accumulation in *Brassica juncea* L. Ecotoxicol Environ Saf 71:76–85

Shaw BP, Sahu SK, Mishra RK (2004) Heavy metal induced oxidative damage in terrestrial plants. In: Prasad MNV (ed) Heavy metal stress in plants: from biomolecules to ecosystems. Narosa Publishing House, New Delhi, India, pp 84–126

Sheng XF, Xia JJ (2006) Improvement of rape (*Brassica napus*) plant growth and cadmium uptake by cadmium resistant bacteria. Chemosphere 64:1036–1042

Sheng XF, Xia JJ, Jiang CY, He LY, Qian M (2008) Characterization of heavy metal-resistant endophytic bacteria from rape (*Brassica napus*) roots and their potential in promoting the growth and lead accumulation of rape. Environ Pollut 156:1164–1170

Sirko A, Brodzik R (2000) Plant ureases: roles and regulation. Acta Biochim Pol 47:1189–1195

Skorzynska-Polit E, Drazkiewicz M, Krupa Z (2010) Lipid peroxidation and antioxidative response in Arabidopsis thaliana exposed to cadmium and copper. Acta Physiol Plant 32:169–175

Sposito FG (2000) The chemistry of soils. In: Maier RM, Pepper IL, Gerba CB (eds) Environmental microbiology. Academic, London, p 406

Sudo E, Itouga M, Yoshida-Hatanaka K, Ono Y, Sakakibara H (2008) Gene expression and sensitivity in response to copper stress in rice leaves. J Exp Bot 59:3465–3474

Suresh B, Ravishankar GA (2004) Phytoremediation- a novel and promising approach for environmental clean-up. Crit Rev Biotechnol 24:97–124

Tak HI, Inam A, Inam A (2010) Effects of urban wastewater on the growth, photosynthesis and yield of chickpea under different levels of nitrogen. Urban Water J 7:187–195

Tank N, Saraf M (2003) Phosphate solubilization, exopolysaccharide production and indole acetic acid secretion by rhizobacteria isolated from *Trigonella graecum*. Indian J Microbiol 43:37–40

Timmis KN, Pieper DH (1999) Bacteria designed for bioremediation. Trends Biotechnol 17:200–204

Tinker PB (1984) The role of microorganisms in mediating and facilitating the uptake of plant nutrients from soil. Plant Soil 76:77–91

Tomsett AB, Thurman DA (1988) Molecular biology of metal tolerances of plants. Plant Cell Environ 11:383–394

Traina SJ, Laperche V (1999) Contaminant bioavailability in soils, sediments, and aquatic environments. Proc Natl Acad Sci U S A 96:3365–3371

Tsavkelova EA, Cherdyntseva TA, Netrusov AI (2005) Auxin production by bacteria associated with orchid roots. Microbiology 74:46–53

Umrania VV (2006) Bioremediation of toxic heavy metals using acidothermophilic autotrophes. Bioresour Technol 97:1237–1242

Van Steveninck RFM, Van Steveninck ME, Wells AJ, Fernando DR (1990) Zinc tolerance and the binding of zinc as zinc phytate in *Lemna minor*. X-ray microanalytical evidence. J Plant Physiol 137:140–146

Vangronsveld J, Clijsters H (1994) Toxic effects of metals. In: Farago MG (ed) Plants and the chemical elements. VHC-Verbgsgesellschaft, Weinheim, Germany, p 149

Vivas A, Biru B, Ruiz-Lozano JM, Azcon R (2006) Two bacterial strains isolated from Zn-polluted soil enhance plant growth and micorrhizal efficiency under Zn toxicity. Chemosphere 52:1523–1533

Vogel JP, Woeste KE, Theologis A, Kieber JJ (1998) Recessive and dominant mutations in the ethylene biosynthetic gene ACS5 of Arabidopsis confer cytokinin insensitivity and ethylene overproduction, respectively. Proc Natl Acad Sci U S A 95:4766–4771

Vogeli-Lange R, Wagner GJ (1990) Subcellular localization of cadmium and cadmium-binding peptides in tobacco leaves: implication of a transport function for cadmium binding peptides. Plant Physiol 92:1086–1093

Wani PA, Khan MS, Zaidi A (2007) Synergistic effects of the inoculation with nitrogen fixing and phosphate solubilizing rhizobacteria on the performance of field grown chickpea. J Plant Nutr Soil Sci 170:283–287

Welch RM (1995) Micronutrient nutrition of plants. Crit Rev Plant Sci 14:49–82

Wenzel WW, Bunkowski M, Puschenreiter M, Horak O (2003) Rhizosphere characteristics of indigenously growing nickel hyperaccumulator and excluder plants on serpentine soil. Environ Pollut 123:131–138

Yang SF, Hoffman NE (1984) Ethylene biosynthesis and its regulation in higher plants. Annu Rev Plant Physiol 35:155–189

Yang J, Kloepper JW, Ryu CM (2009) Rhizosphere bacteria help plants tolerate abiotic stress. Trends Plant Sci 14:1–4

Yu X, Liu X, Zhu TH, Liu GH, Mao C (2012) Co-inoculation with phosphate-solubilizing and nitrogen-fixing bacteria on solubilization of rock phosphate and their effect on growth promotion and nutrient uptake by walnut. Eur J Soil Biol 50:112–117

Zaidi S, Usmani S, Singh BR, Musarrat J (2006) Significance of *Bacillus subtilis* strain SJ 101 as a bioinoculant for concurrent plant growth promotion and nickel accumulation in *Brassica juncea*. Chemosphere 64:991–997

Zhuang XL, Chen J, Shim H, Bai Z (2007) New advances in plant growth-promoting rhizobacteria for bioremediation. Environ Int 33:406–413

Toxicity Reference Values and Tissue Residue Criteria for Protecting Avian Wildlife Exposed to Methylmercury in China

Ruiqing Zhang, Fengchang Wu, Huixian Li, Guanghui Guo, Chenglian Feng, John P. Giesy, and Hong Chang

Contents

1 Introduction 54
2 Data Collection and Analysis 55
   2.1 Selection of Representative Species in China 55
   2.2 Selection of Toxicity Data 56
   2.3 Methods of Deriving TRVs and TRCs 57
3 Review of MeHg Toxicity to Birds 58
4 Derivations of TRVs and TRCs 66
5 Reasonableness of TRVs and TRCs 68
6 Comparison to Ambient Concentrations in Tissues 71
7 Evaluation of Uncertainties 72
8 Summary 74
References 75

R. Zhang
Guangzhou Institute of Geochemistry, Chinese Academy of Science, Guangzhou 510640, China

Graduate University of Chinese Academy of Sciences, Beijing 100049, China

State Key Laboratory of Environmental Criteria and Risk Assessment,
Chinese Research Academy of Environmental Sciences, Beijing 100012, China

F. Wu (✉) • G. Guo • H. Li • C. Feng • H. Chang
State Key Laboratory of Environmental Criteria and Risk Assessment,
Chinese Research Academy of Environmental Sciences, Beijing 100012, China
e-mail: wufengchang@vip.skleg.cn

J.P. Giesy
Department of Veterinary Biomedical Sciences and Toxicology Centre,
University of Saskatchewan, Saskatoon, SK, Canada

Zoology Department and Center for Integrative Toxicology, Michigan State University,
East Lansing, MI 48824, USA

Department of Biology & Chemistry and State Key Laboratory in Marine Pollution,
City University of Hong Kong, Kowloon, Hong Kong, SAR, China

School of Biological Sciences, University of Hong Kong, Hong Kong, SAR, China

1 Introduction

Mercury (Hg) is a globally distributed environmental contaminant with both natural and anthropogenic sources. Of the forms and oxidation states of Hg, the organic form, methylmercury (MeHg), is the most biologically available and the most toxic (Scheuhammer et al. 2007). MeHg can be neurotoxic, embryotoxic, and can impair physiological function, particularly by disrupting endocrines (Tan et al. 2009) and altering reproductive behavior (Frederick and Jayasena 2010). Because MeHg can be bioaccumulated and biomagnified through the food web, diet is the major pathway by which vertebrates are exposed (Liu et al. 2008). Species occupying higher trophic levels in aquatic systems are considered to be at the greatest exposure risk, particularly birds at trophic levels 4 or 5. Although concentrations of Hg can exist in surface water at or near historical background concentrations, the concentrations of Hg that exist in wildlife are higher (Liu et al. 2008). Chronic dietary exposure to relatively small, environmentally relevant concentrations of MeHg is sufficient to be accumulated by tissues to concentrations that impair reproduction of birds (Frederick and Jayasena 2010).

Environmental contamination by Hg released from human activities is a major concern in China (Feng 2005). Concentrations of Hg from anthropogenic emissions that are greater than the historical and regional background levels (Zheng et al. 2010) are extensively distributed, and have been detected in surface waters and tissues of birds (Feng 2005; He et al. 2010; Jin et al. 2006; Zhu et al. 2012). However, no specific guidelines, standards, or criteria have been established for the risk that MeHg may pose to wildlife in China. Assessing the risk that MeHg poses to birds in Chinese aquatic systems is urgently needed to support national policy-making decisions. Thus, derivation of Hg wildlife criteria values that apply to the aquatic systems characteristic of China is a primary task of aquatic environmental managers.

Recently, using the tissue residue approach has been recommended for assessing the ecological risk of bioaccumulative contaminants (Sappington et al. 2011; Beckvar et al. 2005; Newsted et al. 2005). Because wildlife regularly consume fish, toxicity reference values (TRVs) that are based on fish tissue concentrations have been developed for 2,3,7,8-tetrachlodibenzo-*p*-dioxin (TCDD), polychlorinated biphenyls (PCB), perfluorooctane sulfonate (PFOS), and cadmium (Cd) (Newsted et al. 2005; Kannan et al. 2000; Blankenship et al. 2008; Stanton et al. 2010). Moreover, concentrations of contaminants in tissues of wildlife, such as blood and feathers (Kahle and Becker 1999; Herring et al. 2009), have been used as exposure indices for risk assessments. Using the cumulatively ingested dose of a chemical from consuming contaminated food (e.g., tissues) is more accurate in that it accounts for bioaccumulation and bioavailability. A direct relationship between toxicity and the consumed (internal) dose can be either measured or predicted (Sappington et al. 2011). Exposure to the internal dose can be expressed on a tissue-specific or whole-body basis. The internal dose can be either measured or predicted from key ratios, such as bioconcentration, -accumulation, and -magnification factors (BCF, BAF, and BMF), respectively, from trophic magnification factors (TMF), or from more complex pharmacokinetic models. Therefore, TRVs and tissue residue criteria (TRCs) that are based on concentrations of toxicants in tissues are effective for protecting wildlife from the hazards of exposure to pollutants. To illustrate, the critical blood concentration of lead (Pb) has been derived for wildlife by using the tissue residue approach (Buekers et al. 2009), and TRVs were derived for PFOS in avian tissues (e.g., serum and liver) and eggs (Newsted et al. 2005).

Establishing regional criteria is preferred, because species composition, bioaccumulation rates and wildlife diets vary among locations. Canada and the USA have established criteria for assessing potential adverse effects on wildlife from exposure to MeHg. The Canadian Council of Ministers of Environment (CCME) derived wildlife guidelines that were based on concentrations of MeHg in fishes, in which the body mass (bm) and rate of food ingestion by Wilson's storm petrel (*Oceanites oceanicus*) were incorporated (CCME 2000). The U.S. Environmental Protection Agency (US EPA) developed criteria for protecting wildlife that were based on concentrations of MeHg in water, in which the body mass, rate of food ingestion, and BMF for three representative bird species endemic to the North American Great Lakes were used (US EPA 1995b). In addition, by using the Great Lakes Water Quality Initiative (GLWQI), the US EPA developed a set of application factors to account for uncertainties. In these previously developed guidelines and criteria (US EPA 1995b; CCME 2000), the effects of MeHg on reproduction of mallards were used as the critical basis for deriving criteria.

Recently, several new studies of the toxicity of MeHg to birds have become available. The results of these studies suggest that the mallard is not the most sensitive avian species to the effects of MeHg (Heinz et al. 2009, 2010a), and thus may not be representative or protective of other species. Therefore, it was deemed desirable to update the TRV values to reflect the effects of MeHg on birds that consume aquatic biota. Such an update can then be applied to representative species endemic to China. The latest research on the toxicity of MeHg to birds was reviewed, and thresholds of toxicity were derived that were based on concentrations of MeHg in the diet (fish) and on concentrations of MeHg in tissues of birds. Finally, estimates of TRV and TRC values to protect birds in Chinese aquatic systems from the effects of MeHg were developed.

2 Data Collection and Analysis

2.1 Selection of Representative Species in China

According to viewpoints expressed in the technical support document for the GLWQI, the primary basis for selecting representative avian species is exposure to contaminants through aquatic food chains, such as fish-consuming species. The species that experience the greatest exposure are favored as representative avian species (US EPA 1995c). Three species that commonly inhabit Chinese aquatic ecosystems

Table 1 Body mass and rate of food ingestion for three representative species

Species and life history parameters	Value	References
Night heron (*Nycticorax nycticorax*)		
Body mass (kg)	0.706[a, b]	Dunning (1993)
Rate of ingestion (kg/day)	0.239[c]	
Little egret (*Egretta garzetta*)		
Body mass (kg)	0.342[a]	Zamani-Ahmadmahmoodi
Rate of ingestion (kg/day)	0.148[c]	et al. (2010); Fujita (2003); Zhang and Liu (1991)
Eurasian spoonbill (*Platalea leucorodia*)		
Body mass (kg)	2.232	Liu et al. (2003)
Rate of ingestion (kg/day)	0.514[c]	

[a]Geometric mean of the data from different studies
[b]Geometric mean of the values reported (Dunning 1993) and the data from China Digital Science and Technology Museum (http://amuseum.cdstm.cn/AMuseum/dongwu/page/animal_detail_4683.html)
[c]Calculated from the allometric equation (Nagy 2001): $FI = 3.048 \times BW^{0.665}$

were selected as representative species in China. These three species were the night heron (*Nycticorax nycticorax*), little egret (*Egretta garzetta*), and Eurasian spoonbill (*Platalea leucorodia*), all of which consume aquatic prey. These three species are widely distributed in Chinese aquatic ecosystems (Barter et al. 2005), and each has been studied extensively as indicators of environmental pollution and wetlands' health (Zamani-Ahmadmahmoodi et al. 2010; Burger and Gochfeld 1997; Zhang et al. 2006; An et al. 2006). The night heron and Eurasian spoonbill are species regarded as second-grade state-protected animals in China. The little egret and Eurasian spoonbill are species listed in the Convention on International Trade in Endangered Species of Wild Fauna and Flora. Body mass and rates of food ingestion for these three species are summarized in Table 1.

2.2 Selection of Toxicity Data

Information on effects of MeHg on birds has been summarized by the US EPA (US EPA 1995b). Toxicity threshold values for MeHg, expressed as no observed adverse effects levels (NOAEL) or lowest observed adverse effects levels (LOAEL), were derived from several endpoints, and were determined for avian wildlife based on concentrations ingested by eating fish and bird tissues. Dietary-based data were converted to average daily intake (ADI) values and were expressed as units of μg MeHg/g body mass/day (μg MeHg/g (bm)/day). ADI values were calculated from body masses and rates of ingestion by the selected surrogate birds. When rates of food ingestion were not reported in a paper, they were calculated by using the most recent allometric equations (Nagy 2001).

The principles used as the basis for selecting utilizable NOAEL or LOAEL values were as follows (CCME 1998): (1) the study retained suitable control conditions; (2) the study was designed to consider ecologically relevant endpoints, such as reproduction, embryonic development, offspring or survival of adults (F_0), growth, and other responses; (3) a clear dose–response relationship was demonstrated in the study; (4) the form and dosage of test chemical were reported; (5) the tested chemical was administered via the oral, rather than by other routes (i.e., only the oral route is natural for wildlife in the field); and (6) studies that included only acute exposures were not accepted, because they provided no data on chronic, sublethal effects on wildlife.

2.3 Methods of Deriving TRVs and TRCs

Two methods were used to derive TRVs from dietary or tissue concentrations. These were the critical study approach (CSA) and the species sensitivity distribution (SSD) approach. TRVs that were based on dietary exposure are expressed as daily dietary intake (μg MeHg/g (bm)/day). TRVs that were based on dietary exposure were converted to the corresponding dietary-based TRC values by using body masses and rates of food ingestion by the three representative surrogate species. The TRVs that were based on concentrations of MeHg in tissues of birds do not vary among representative species as a function of body mass and rate of ingestion.

CSA. CSA is the primary method for assessing risk to wildlife and for deriving criteria for protection of wildlife (CCME 1998; US EPA 1995a, b, 2003, 2005; Sample and Suter 1993). This method is used to select the critical study for deriving recommended TRVs, which involves finding a technically defensible, definitive study from which a toxicity threshold is bracketed by experimental doses (Blankenship et al. 2008; US EPA 2003). A series of uncertainty factors (UFs) are applied to LOAEL or NOAEL values that are obtained from the critical study, and these are used to determine the TRVs. A UF is assigned from guidance given in the Technical Support Document (TSD) for Wildlife Criteria for the GLWQI (US EPA 1995c), and in the GLWQI Criteria Documents for the Protection of Wildlife (US EPA 1995b). Three sources of uncertainty are considered in assigning a UF value: (1) interspecies differences in toxicological sensitivity (UF_A), (2) subchronic to chronic extrapolations (UF_S), and (3) LOAEL to NOAEL extrapolations (UF_L). Application factors for each source of uncertainty were assigned values between 1 and 10, based on available information and professional judgment (US EPA 1995c; Newsted et al. 2005).

SSD. SSD is a statistical distribution representing the variation in sensitivity of species to a contaminant, and can be developed by a statistical or empirical distribution function of response for a sample of species (Posthuma et al. 2002). This method has been used to assess risks to aquatic organisms and for deriving water quality criteria (WQC) for protecting aquatic species (Caldwell et al. 2008; Hall et al. 1998; Solomon et al. 1996; Stephan et al. 1985). However, because data on the toxicity of contaminants to wildlife are often insufficient, the SSD approach to assess wildlife risks has not often been applied. For the analysis presented herein, the SSD was used to determine the concentration of MeHg that would be protective of wildlife. This concentration is the fifth centile (HC_5) of the SSD generated from selected wildlife effects data for MeHg. Data representing the most sensitive endpoint for each species were selected to construct the SSD. If the duration of exposure was deemed to be insufficient, or if only an unbounded LOAEL was produced in some studies, the data were corrected before fitting the SSD function, by using UF_S or UF_L (US EPA 2005). The basic assumption of the SSD approach is that sensitivity among species can be described by using a specified statistical distribution, such as the normal (Wagner and Løkke 1991; Aldenberg and Jaworska 2000), logistic (Kooijman 1987; Aldenberg and Slob 1993), triangular (Stephan et al. 1985), or Weibull (Caldwell et al. 2008) probability functions, or by using distribution-free, nonparametric methods (Ling 2004; Newman et al. 2000). It was assumed that the toxicity data selected for MeHg are skewed and can be described by using a log-normal distribution (Aldenberg and Jaworska 2000). The ETX2.0 program was used to fit the distribution (Van Vlaardingen et al. 2004). The HC_5 and its two-sided 90% confidence limits, designated as lower limit (LL HC_5) and upper limit (UL HC_5), were calculated. The goodness of fit was tested with the Anderson-Darling and Kolmogorov-Smirnov tests to ensure that the data were log-normally distributed.

3 Review of MeHg Toxicity to Birds

The results of subchronic and chronic toxicity testing on the mallard (*Anas platyrhynchos*), white leghorn chicken (*Gallus domesticus*), ring-necked pheasant (*Phasianus colchicus*), Japanese quail (*Coturnix japonica*), red-tailed hawk (*Buteo jamaicensis*), and zebra finch (*Poephila quttata*) have been summarized in the GLWQI Criteria Documents for the Protection of Wildlife (US EPA 1995b). No additional toxicity information was available for these species except for mallard. Thus, the toxicity threshold concentrations from diet and tissue data for the other five species were used directly (Table 2). Below, a review of the recent relevant toxicity studies is presented.

Mallard (A. platyrhynchos). As stated in the GLWQI Criteria Documents (US EPA 1995b), the dietary LOAEL and NOAEL values for mortality and for neurological effects on the mallard were 3.0 and 0.5 μg MeHg/g feed (dry weight, dwt), respectively (Heinz and Locke 1976). Using the average body mass of 1 kg for a mallard (Delnicki and Reinecke 1986), a rate of food ingestion of 0.051 kg dried feed/kg fresh (bm)/day was derived from Nagy's (Nagy 2001) allometric equation for omnivorous birds, rather than the equation (Nagy 1987) that was used in the GLWQI Criteria Documents for Protection of Wildlife. Assuming that the laboratory feed for the mallard consists of 10% water (US EPA 1995b), the rate of food ingestion would be equivalent to 0.057 kg food (wwt)/kg (bm)/day. Corresponding values for LOAEL and NOAEL were calculated to be 0.171 and 0.029 μg MeHg/g (bm)/day. The LOAEL and NOAEL values, expressed as concentrations of MeHg in egg, were, respectively, 0.79 and 5.64 μg MeHg/g (wwt) (Heinz and Locke 1976). Based on multigenerational effects on reproduction, when converted from the dietary values by the use of rates of consumption of food of 0.156 kg food (wwt)/kg (bm)/day (US EPA 1995b), the dietary LOAEL value was 0.078 μg MeHg/g (bm)/day.

Table 2 Summary of subchronic and chronic avian toxicity data for MeHg (μg MeHg/g (bm)/day for ADI; μg MeHg/g (wwt) for diet or tissues)

Species	Media	Toxic effects observed	NOAEL	LOAEL	References
Mallard	Diet	Offspring mortality and neurotoxicity	0.5	3.0	Heinz (1974, 1975, 1976a, b, 1979)
	ADI	Offspring mortality and neurotoxicity	0.029	0.171	
	Egg	Offspring mortality and neurotoxicity	0.79	5.64	
	Diet	Multigenerational exposure, reproduction		0.5	
	ADI	Multigenerational exposure, reproduction		0.078	
	Egg	Multigenerational exposure, reproduction		0.83	
	Liver	Multigenerational exposure, reproduction		1.29	
	Kidney	Multigenerational exposure, reproduction		1.64	
	Breast muscle	Multigenerational exposure, reproduction		0.77	
	Brain	Multigenerational exposure, reproduction		0.55	
	Ovary	Multigenerational exposure, reproduction		0.58	
	Primary feather	Multigenerational exposure, reproduction		9.71	
	Diet	Reproduction, duckling survival	2	4	Heinz et al. (2010a)
	ADI	Reproduction, duckling survival	0.114	0.228	
	Egg	Reproduction, duckling survival	3.7	5.9	
Great egret	Diet	Behavior, growth, biochemistry		0.5	Bouton et al. (1999); Hoffman et al. (2005); Spalding et al. (2000a); Spalding et al. (2000b)
	ADI	Behavior, growth, biochemistry		0.091	
	Blood	Behavior, growth, biochemistry		10.3	
	Liver	Behavior, growth, biochemistry		15.1	
	Brain	Behavior, growth, biochemistry		3.4	
	Kidney	Behavior, growth, biochemistry		8.1	

(continued)

Table 2 (continued)

Species	Media	Toxic effects observed	NOAEL	LOAEL	References
Common loon	Diet	Growth	1.5		Kenow et al. (2003)
	ADI	Growth	0.27		
	Blood	Growth	3.33 µg/mL		
	Diet	Immune function, biochemistry	0.08	0.4	Kenow et al. (2008); Kenow et al. (2007a); Kenow et al. (2007b)
	ADI	Immune function, biochemistry	0.014	0.072	
	Blood	Immune function, biochemistry		1.98	
	Brain	Immune function, biochemistry		0.88	
	Kidney	Immune function, biochemistry		2.29	
	Breast muscle	Immune function, biochemistry		1.23	
	Liver	Immune function, biochemistry		4.03	
	Feather	Immune function, biochemistry		22.03	
American kestrel	Diet	Reproduction		0.3	Albers et al. (2007)
	ADI	Reproduction		0.055	
	Egg	Reproduction		2.00[a]	
	Diet	Immune function		0.23	Fallacara et al. (2011)
	ADI	Immune function		0.043	
	Blood	Immune function		6.12[a]	
	Spleen	Immune function		5.16[a]	
White ibis	Diet	Reproduction		0.05	Frederick and Jayasena (2010)
	ADI	Reproduction		0.010	
	Feather	Reproduction		6.32[a]	
	Blood	Reproduction		0.73[a]	
White leghorn chicken	ADI	Growth		0.29	Fimreite (1970)
	Liver	Growth		3.49	
	ADI	Mortality	0.57	0.86	
	Liver	Mortality	7.25	10.00	
	ADI	Reproduction		0.67	Scott (1977)

Ring-necked pheasant	ADI	Mortality	0.25		Spann et al. (1972)
	ADI	Reproduction		0.75	
	ADI	Reproduction		0.25	
Japanese quail	ADI	Offspring mortality		0.093	Fimreite (1971)
Red-tailed hawk	ADI	Neurological effect and mortality	0.26	0.52	Eskeland and Nafstad (1978)
Zebra finch	ADI	Neurological effect and mortality	0.49	1.2	Fimreite and Karstad (1971)
	ADI	Neurological effect and mortality	0.88	1.75	Scheuhammer (1988)
	Liver	Neurological effect and mortality		30.5–73	
	Kidney	Neurological effect and mortality		35.5–65	
	Brain	Neurological effect and mortality	7–11	14.1–20	

[a]Total mercury

NOAEL no observed adverse effects level, *LOAEL* lowest observed adverse effects level, *ADI* average daily intake, *bm* body mass, *wwt* wet weight

The corresponding concentrations of MeHg in tissues such as liver, kidney, breast muscle, brain, ovary, and primary feathers, and in eggs are given (geometric mean for three generations) for mallards (Heinz 1979; Table 2).

Reproductive effects of MeHg on mallards were investigated by exposing adults to one of four doses of MeHg (1, 2, 4, or 8 μg MeHg/g (dwt)) (Heinz et al. 2010a). No adverse effects were observed in adults, or on egg fertility or the rate of hatching success. However, at doses of 4 or 8 μg MeHg/g (dwt), survival of ducklings and the number of ducklings produced per female were less than those of untreated controls. Ducklings at 6 days of age from parents fed 4 or 8 μg MeHg/g (dwt) weighed less than the controls. Thus, doses of 2 and 4 μg MeHg/g (dwt) were considered to be the dietary-based NOAEL and LOAEL values, respectively. However, both doses were greater than the LOAEL of 0.5 μg MeHg/g (dwt) (Heinz 1979), which was determined from the results of a multigeneration exposure. The corresponding daily doses, which were converted from the dietary values using a rate of ingestion of food of 0.057 kg/kg (bm)/day, were 0.114 and 0.228 μg MeHg/g (bm)/day, respectively. The NOAEL and LOAEL for Hg concentrations in eggs are provided in Table 2.

When lesser doses of MeHg were injected into mallard eggs, hormesis was observed at 0.05 μg MeHg/g (wwt) (least dose) (Heinz et al. 2011), which agrees with a similar observation, in which mallards were exposed to a single dose of 0.5 μg MeHg/g (dwt) (Heinz et al. 2010b). However, the mean concentration of MeHg in eggs of hens fed MeHg in the diet was 0.81 μg MeHg/g (wwt) (Heinz et al. 2010b), a value greater than the 0.05 μg/g (wwt) of Hg that was injected. The exposure pathway was regarded to be the major reason for the difference. It was suggested that when MeHg was injected into embryos it was more toxic than equivalent concentrations of maternally deposited MeHg (Heinz et al. 2009, 2011). Although the mechanism for this discrepancy is not well understood, it is probably due to a difference in dose vs. dose rate, or from the binding of MeHg, which results from differences in biological activity between the two vectors of exposure. The results of a single dose MeHg exposure (0.5 μg MeHg/g (dwt)) (Heinz et al. 2010b) seemed to contradict the results of the multigeneration, dietary exposure in which the same dose was used. This might result from differences between forms of MeHg and the sources of mallards tested (Heinz et al. 2010b). Thus, additional research is needed on "low-dose" effects of MeHg to the mallard. According to the analysis above, the results from the multigeneration exposure, rather than those from the latest studies (Heinz et al. 2010a, b), should be used as the basis for developing TRVs and TRCs.

Great egret (Ardea alba). Several different effects, including behavior (Bouton et al. 1999), survival, growth and accumulation in tissues (Spalding et al. 2000a), histology, neurology, and immunology (Spalding et al. 2000b) from exposure to MeHg, were addressed in three sladies, in which juvenile great egrets were exposed to two dietary doses of 0.5 or 5 μg MeHg/g (wwt) for 12 weeks. Severe ataxia was observed in individuals fed the greater dose, whereas the lesser dose produced effects on activity, a tendency to seek shade, and motivation to hunt prey (Bouton et al. 1999). After 9 weeks of exposure, appetite and mass declined significantly in both dosed groups (Spalding et al. 2000a). Adverse effects, related to immune function, were observed in individuals fed the lesser dose, whereas individuals fed the greater dose exhibited adverse effects on tissues related to immune and nerve functions

(Spalding et al. 2000b). In a study of the biochemical effects of MeHg (Hoffman et al. 2005), only activities of the enzyme aspartate aminotransferase (AST) in blood plasma, and thiobarbituric acid-reactive substances (TBARS) in liver of individuals fed the lesser dose were significantly greater than those of the control group. However, there were significant changes in activities of glutathione peroxidase (GSH-Px), aspartate aminotransferase (AST), lactate dehydrogenase, and in the concentrations of uric acid, total protein, and inorganic phosphorus in tissues of individuals fed the greater dose. Therefore, on the basis of effects observed, the dietary LOAEL value for the great egret was determined to be 0.5 µg MeHg/g (wwt). The corresponding LOAEL values were based on concentrations in plasma, liver, brain, and kidney, and are presented in Table 2. Using an average bm of 1.0 kg for great egret reported by Rumbold et al. (2008), and a rate of intake of food of 0.181 kg/day estimated from an allometric equation for wading birds (Kushlan 1978; US EPA 1993), the dietary LOAEL was calculated to be 0.091 µg MeHg/g (bm)/day.

Common loon (Gavia immer). The effects of MeHg on growth (Kenow et al. 2003), behavior (Kenow et al. 2010), immune function (Kenow et al. 2007a), and the biochemical index (Kenow et al. 2008) of juvenile common loons were investigated during which individuals were exposed for 15 weeks. Neither adverse effects on growth or survival occurred in loons that were exposed to three doses, 0.1, 0.5, or 1.5 µg MeHg/g (wwt) (Kenow et al. 2003), nor behavioral effects at doses of 0.08, 0.4, or 1.2 µg MeHg/g (wwt) (Kenow et al. 2010). However, adverse effects on immune function (Kenow et al. 2007a) and effects related to oxidative stress and altered glutathione metabolism (Kenow et al. 2008) occurred at 0.4 and 1.2 µg MeHg/g (wwt). Using a bm of 4.67 kg for common loon adults (Barr 1996; Dunning 1993), a food intake rate of 0.839 kg/day was derived from the allometric equation for carnivorous birds (Nagy 2001). Thus, the dietary-based NOAEL value, based on growth effects of the common loon, was 0.27 µg MeHg/g (bm)/day (1.5 µg MeHg/g (wwt)). The NOAEL and LOAEL values were based on dietary exposure and effects on immune function and biochemistry, and were, respectively, 0.014 µg MeHg/g (bm)/day (0.08 µg MeHg/g (wwt)) and 0.072 µg MeHg/g (bm)/day (0.4 µg MeHg/g (wwt)). The corresponding concentration of MeHg in blood that was associated with growth effects was 3.33 µg MeHg/mL (Kenow et al. 2003). In Table 2, we show the LOAEL values for blood, brain, kidney, breast muscle, liver, and feathers (geometric mean of Hg in feather at different sites); these were based on immune and biochemical-function effects. The corresponding NOAEL values were not available (Kenow et al. 2007b).

White ibis (Eudocimus albus). Juvenile white ibises were exposed to dietary MeHg at three doses of 0.05, 0.1, or 0.3 µg MeHg/g (wwt), and their foraging behavior and efficiency (Adams and Frederick 2008), survival (Frederick et al. 2011), and breeding behavior (Frederick and Jayasena 2010) were examined. Hormetic effects were observed on foraging efficiency at doses of 0.05 and 0.1 µg MeHg/g (wwt), and the effects from exposure to the 0.3 µg MeHg/g (wwt) group were similar to that of the controls. However, no clear dose–response relationship was demonstrated in the Adams and Frederick (2008) study. Therefore, we did not further utilize this study for deriving TRVs. Exposure to MeHg at these three doses did not affect the survival of white ibises (Frederick et al. 2011). The effect on breeding behavior was investigated by exposing white ibises to these three doses of MeHg over a 3-year period (Frederick and Jayasena 2010). The effects produced were increases in male–male pairing behavior, dose-related reductions in key courtship behaviors for males, and fewer eggs laid (Frederick and Jayasena 2010). In addition, productivity per nest by heterosexual males was significantly less than that of controls. These reproductive effects could be related to changes in endocrine function, because concentrations of estradiol and testosterone were altered in birds exposed to all three doses of MeHg (0.05, 0.1 and 0.3 μg MeHg/g (wwt)) (Jayasena et al. 2011). In another study, endocrine function of white ibises was affected by relatively small concentrations of MeHg. These small exposures might have changed reproductive behavior and altered mate choice in males, and may have reduced reproductive success so as to finally influence population numbers. Studies of wild birds (Heath et al. 2005) also suggest that exposure to Hg may result in fewer birds nesting or more nest abandonment from subacute effects on hormone systems. From these studies, the dietary LOAEL value for a reproductive endpoint in white ibises exposed to MeHg was established to be 0.05 μg MeHg/g (wwt), and the corresponding Hg tissue residue concentrations were 6.32 μg total Hg (THg)/g (wwt) (geometric mean of values during 2006 and 2008) in blood, and 0.73 μg THg/g (wwt) in feathers (Frederick and Jayasena 2010). Concentrations of THg in blood and feathers were derived primarily from MeHg food residue information. Using a white ibis bm of 0.869 kg (geometric mean) (Dunning 1993), and a food ingestion rate of 0.182 kg/day (21% of bm) (Kushlan 1977), the corresponding LOAEL was derived to be 0.010 μg MeHg/g (bm)/day.

American Kestrel (Falco sparverius). American kestrels were exposed to 0.3, 0.7, 1.2, 1.7, or 2.2 μg MeHg/g (wwt), and parameters of their reproduction were measured (Albers et al. 2007). Adverse effects on reproduction were observed in birds at all doses. At a dose of 0.3 μg MeHg/g (wwt), the number of fledglings and the percent of nestlings fledged were lesser than that of controls and were lesser yet at the greater doses (i.e., 0.7 or 1.2 μg MeHg/g (wwt)). Total failure of fledging was observed at 1.7 μg MeHg/g (wwt). Thus, 0.3 μg MeHg/g (wwt) was considered to be a LOAEL for reproductive effects in American kestrels. The LOAEL, based on concentrations in eggs, was 2.00 μg MeHg/g (wwt). Using a bm of 0.119 kg and a food ingestion rate of 0.022 kg/day (geometric means) (Dunning 1993; Yáñez et al. 1980; US EPA 1993), a dietary-based LOAEL value of 0.055 μg MeHg/g (bm)/day was calculated.

Effects of MeHg on immune function and hematology were determined for adult male American kestrels exposed to 0.23 or 1.5 μg MeHg/g (wwt) for a period of 13 weeks (Fallacara et al. 2011). Suppression of immune function occurred at both doses. Adult kestrels were more sensitive to the effects of MeHg for immune function than for reproduction. The 0.23 μg MeHg/g (wwt) dose was assigned as the dietary LOAEL value for immunotoxic effects. Using bm and rate of ingestion of food, this LOAEL was converted to an ADI of 0.043 μg MeHg/g (bm)/day. The corresponding concentrations of total Hg in blood and spleen after exposure are shown in Table 2. The toxic effects observed in the foregoing studies for birds are summarized in Table 2, and the LOAEL and NOAEL values, expressed as ADI, are shown in Fig. 1. Reproduction, immune-function, and effects on biochemistry were more sensitive to MeHg exposure than were behavior, mortality, and neuropathology effects. The reproductive, immune

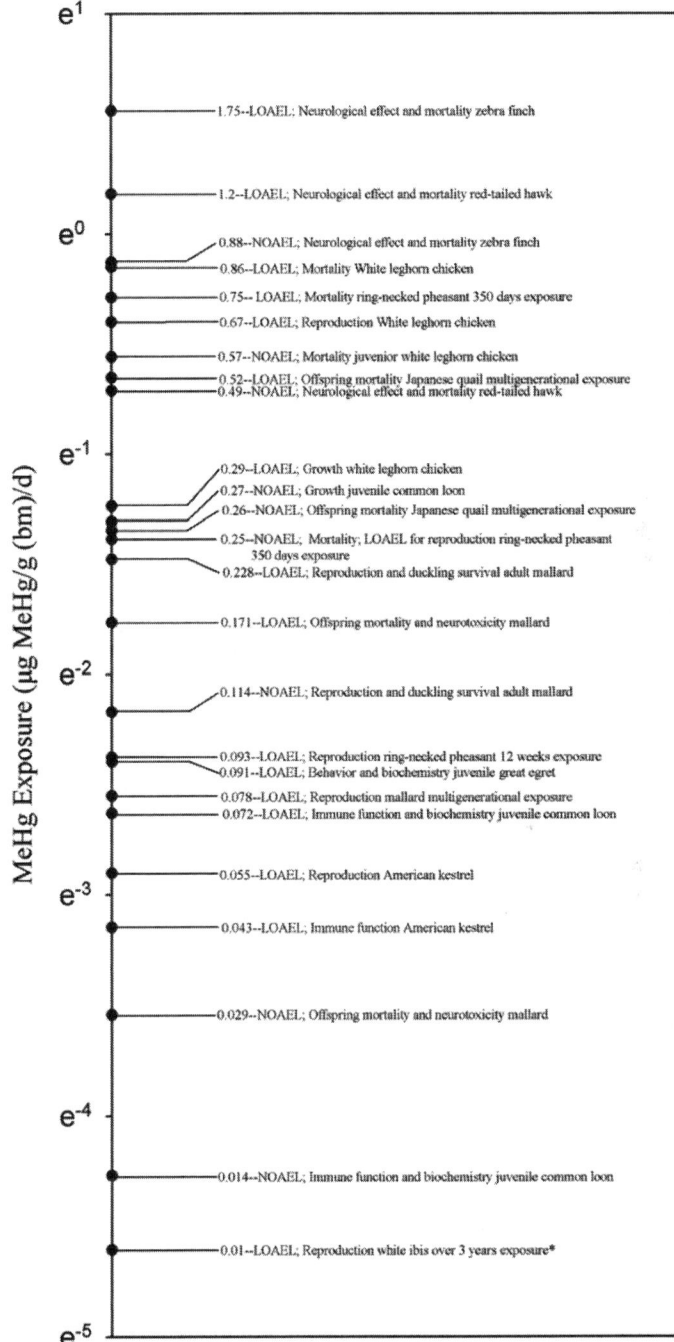

Fig. 1 Toxicity thresholds for avian species dietary exposure to MeHg expressed as the average daily intake (ADI). The *asterisk* indicates that the data are critical for deriving the criteria values. *NOAEL* no observed adverse effects level, *LOAEL* lowest observed adverse effects level, *bm* body mass. See Table 2 for data set

function, and biochemistry effects occurred over a dose range of 0.01–0.67 μg MeHg/g (bm)/day. The threshold for other adverse effects was 0.029 μg MeHg/g (bm)/day. White ibises, juvenile common loons, mallards, and American kestrels are species that are relatively more sensitive than other species. Although all effects were concentration dependent, the endpoint that was most biologically relevant and sensitivity to it varied among bird species. The most sensitive endpoint effect for MeHg exposure in any species was reproductive productivity of the white ibis (Frederick and Jayasena 2010).

4 Derivations of TRVs and TRCs

CSA. As indicated above, the study on the reproductive effects of MeHg in white ibises (Frederick and Jayasena 2010) was the most appropriate critical study, and the results of that study were used for deriving the TRV and TRC. Both juveniles and adults were exposed over 3 years and three breeding seasons in the critical study. LOAEL values were based on reproductive effects in white ibis, and in particular on male–male pairing behavior and on fewer eggs laid. The LOAEL values were expressed as both dietary (ADI) and residue concentrations in feathers and blood (viz., respectively, 0.010 μg MeHg/g (bm)/day, 6.32 μg THg/g (wwt), and 0.73 μg THg/g (wwt)). Based on reproductive effects of MeHg on white ibises and on characteristics of avian predators, three uncertainty factors were considered as follows: (1) an interspecies uncertainty factor, (2) a LOAEL to NOAEL uncertainty factor, and (3) a subchronic to chronic uncertainty factor. An overall uncertainty factor of 2 was assigned to account for data gaps (US EPA 1995b, c; Table 3).

Table 3 Assignment of uncertainty factors for derivation of avian wildlife toxicity reference values (TRVs) for MeHg

Uncertainty factors	Notes
Interspecies uncertainty factor (UF_A)	The data selected to determine avian toxicity reference values were from a reproductive study of white ibis, which is a piscivorous wading bird in the same order with the three representative birds. Because the white ibis is the most sensitive of the species reviewed in the present study, $UF_A = 1.0$
LOAEL to NOAEL (UF_L)	The data from the critical study is a LOAEL based on a reproduction endpoint in a multiple year exposure, but not a NOAEL. However, the difference between the LOAEL and control was only 13.2% for the decreases in egg productivity. Taken together with the ratio of LOAEL and NOAEL in other studies, the $UF_L = 2.0$
Subchronic to chronic uncertainty (UF_S)	The critical study was conducted over 3 years, covered juvenile stage and 3 breeding seasons, evaluated reproductive behavior and productivity, which are considered ecologically most relevant, thus $UF_S = 1.0$
Overall uncertainty factor (UF) for TRVs	$UF = 1 \times 2 \times 1 = 2$

NOAEL no observed adverse effects level, *LOAEL* lowest observed adverse effects level

Table 4 MeHg toxicity reference values (TRVs) and tissue residue criteria (TRCs) for representative avian species based on average daily intake (ADI) or diet, feathers, and blood

	LOAEL	TRV	TRC
ADI, ng MeHg/g (wwt) for TRC, ng MeHg/g (bm)/day for LOAEL and TRV	10	5.0	15.47
Feather, µg THg/g (wwt)	6.32	3.16	3.16
Blood, µg THg/g (wwt)	0.73	0.365	0.365

LOAEL lowest observed adverse effects level, *bm* body mass, *wwt* wet weight

Table 5 Toxicity thresholds for species sensitivity distribution curve fitting (µg MeHg/g (bm)/day)

Species	Exposure period	Reported value LOAEL	Reported value NOAEL	Converted NOAEL	References
Mallard	3 years	0.078		0.039[a]	Heinz (1974, 1975, 1976a, b, 1979)
Great egret	12 weeks	0.091		0.0455[a]	Bouton et al. (1999), Hoffman et al. (2005), Spalding et al. (2000a), Spalding et al. (2000b)
Common loon	15 weeks		0.014	0.014	Kenow et al. (2008), Kenow et al. (2007a), Kenow et al. (2007b)
American kestrel	13 weeks	0.043		0.0215[a]	Fallacara et al. (2011)
White ibis	3 years	0.01		0.005[a]	Frederick and Jayasena (2010)
White leghorn chicken	21 days	0.29		0.0145[a,b]	Fimreite (1970)
Ring-necked pheasant	12 weeks	0.093		0.0465[a]	Fimreite (1971)
Japanese quail	6 weeks		0.26	0.26	Eskeland and Nafstad (1978)
Red-tailed hawk	12 weeks		0.49	0.49	Fimreite and Karstad (1971)
Zebra finch	76 days		0.88	0.88	Scheuhammer (1988)

[a]The value derived by applying a LOAEL to NOAEL factor of 2.0
[b]The value derived by applying a factor of 10 to account for the uncertainty in establishing a NOAEL from short time (<1 month) (Jongbloed et al. 1996)
NOAEL no observed adverse effects level, *LOAEL* lowest observed adverse effects level, *bm* body mass

TRVs based on ADI and on concentrations of MeHg in feathers and blood were 5.0 ng MeHg/g (bm)/day, 3.16 µg THg/g (wwt), and 0.365 µg THg/g (wwt), respectively (Table 4). Based on the dietary exposure of the representative avian species, the TRC values for the night heron, little egret, and Eurasian spoonbill were 14.77, 11.55, and 21.71 ng MeHg/g (wwt), respectively; the geometric mean was 15.47 ng MeHg/g (wwt). TRCs, based on concentrations of THg in feathers and blood were the same as the corresponding TRVs (Table 4).

SSD. NOAEL and LOAEL values selected for constructing the SSD were corrected by applying a LOAEL to NOAEL factor, or a subchronic to chronic factor (Table 5). On the basis of the function describing the toxicity thresholds of all species (Table 5),

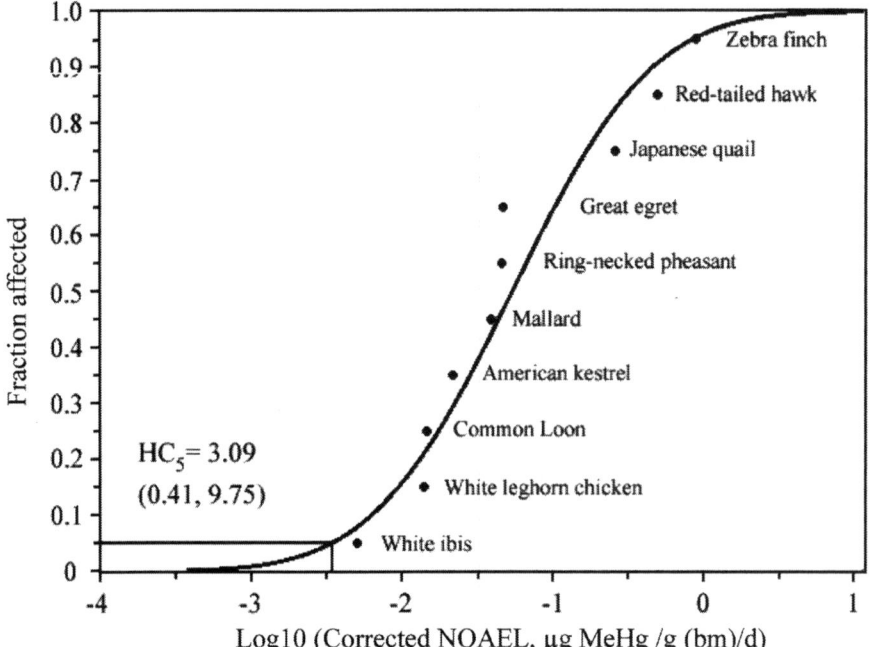

Fig. 2 The distribution of species sensitivity for the avian toxicity data of MeHg. Fifth centile of species sensitivity distribution (HC_5, ng MeHg/g (bm)/day) is presented with two-sided 90% confidence limits given in *parenthesis*. The goodness-of-fit test by the use of the Anderson-Darling and Kolmogorov-Smirnov tests was accepted. The *black dots* represent the toxicity data, and the *solid curve* represents the log-normal distribution. *NOAEL* no observed adverse effects level, *bm* body mass. See Table 5 for data set

the HC_5 was predicted to be 3.09 ng MeHg/g (bm)/day, which was defined as the dietary-based TRV (Fig. 2). Based on dietary exposure, TRCs for the night heron, little egret, and Eurasian spoonbill were 9.13, 7.14, and 13.42 ng MeHg/g (wwt), respectively, and the geometric mean was 9.56 ng MeHg/g (wwt).

5 Reasonableness of TRVs and TRCs

Deriving TRVs and TRCs for birds is always limited by the paucity of toxicological study results. To assess the risk posed by MeHg to birds in China, the TRVs and TRCs, based on concentrations of MeHg in tissues of fish and birds, were derived by applying two methods and incorporating the most recent toxicological data. These criteria values provide points of reference for concentrations of MeHg in aquatic life and birds, and can be used directly in the tissue residue approach to ecological risk assessment. To judge the reasonableness of protective guidelines derived by the two methods, they were compared to criteria developed by others.

CSA. In the criteria documents of the US EPA (US EPA 1995b) and CCME (CCME 2000), effects on reproduction of mallard (Heinz 1974, 1975, 1976a, b, 1979) were used as the most appropriate results for deriving wildlife criteria values. The sensitivity of the reproductive endpoint of mallard to MeHg was similar to that of the reproductive and immune endpoints of American kestrels, but less than that of the immune and biochemical endpoints of juvenile common loons, and the reproductive endpoint of white ibises (Fig. 1). Thus, it was concluded that reproduction of the mallard is relatively insensitive to the effects of MeHg (Heinz et al. 2010a). This conclusion was confirmed by using a study in which the eggs of 26 bird species were injected with MeHg (Heinz et al. 2009). Only one species, the double-crested cormorant (*Phalacrocorax auritus*), was less sensitive than the mallard (Heinz et al. 2009). Therefore, basing the TRVs and TRCs on the reproduction study results in white ibises provides more protection to avian wildlife than basing them only on the results of MeHg effects on the mallard.

A dietary-based TRC of 33 ng MeHg/g (wwt) has been derived by CCME for Wilson's storm petrel from study results on mallard without applying UFs (CCME 2000). If a UF of 2.0 is applied to this value, the TRC for Wilson's storm petrel would be 16.5 ng MeHg/g (wwt), which is similar to the value derived in the present evaluation. However, the TRV given by the CCME would become 15.5 if a UF of 2.0 is applied. This value is three times greater than the TRV of 5.0 ng MeHg/g (bm)/day derived in the present assessment. The difference is because Wilson's storm petrel has a greater ingestion of food (FI):BM ratio of 0.94. The species with the greatest FI:BW ratio has the greatest potential exposure to contaminants. Hence, the selection of representative species is important for deriving a TRC.

A NOAEL of 0.014 µg MeHg/g (bm)/day for effects on immune function and other biochemical effects in juvenile common loons is similar to the value derived for the white ibis. When this value was used to derive a TRV, and a UF_L of 1.0, a UF_S of 1.0 and a UF_A of 3.0 were considered; the resulting avian dietary-based TRV would be 4.7 ng MeHg/g (bm)/day, which is similar to the value derived from the MeHg toxicity to white ibis. The UF_L was set to 1.0 because the common loons study provided a NOAEL rather than a LOAEL (Kenow et al. 2007a). According to pharmacokinetic studies of absorption and elimination of MeHg in common loon chicks (Fournier et al. 2002), the result of a study of common loons, with a duration of 15 weeks, was accepted as being a chronic exposure study. Thus, a UF_S of >1.0 was deemed to be unnecessary (Kenow et al. 2007a). Although the study by Kenow et al. (2007a) did not provide information on the most ecologically relevant endpoint, which is reproduction, the common loon is a piscivorous wading bird and is the second most sensitive avian species to MeHg. Applying a UF_A of 10 is likely to be overly conservative. Therefore, according to the UF_A used in GLWQI Criteria Document (US EPA 1995b), an intermediate value of 3.0 was used for the UF_A.

SSD. SSD is an effective method to represent the variation in sensitivity to chemicals among species. A specified effect level, such as the proportion of species expected to respond to a particular exposure for a specific measurement endpoint, can be determined so that most species are protected. SSDs have been used to assess risk

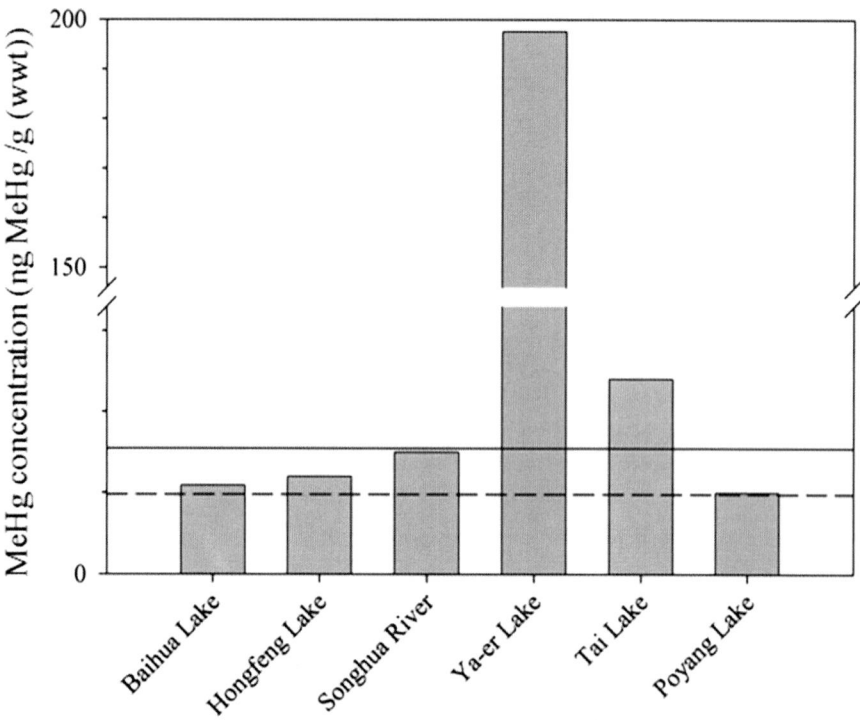

Fig. 3 MeHg concentrations found in fish from different aquatic systems in China, compared to the dietary-based TRC values from critical study approach (CSA) and species sensitivity distribution (SSD) approach. The *solid* and *dashed lines* represent the TRC values from CSA and SSD, respectively. *wwt* wet weight. Baihua Lake (Yan et al. 2008), Hongfeng Lake (He et al. 2010), Songhua River (Zhu et al. 2012), Ya-er Lake (Jin et al. 2006) Tai Lake, and Poyang Lake (Zhang et al. 2006)

and develop water quality criteria for aquatic species (Caldwell et al. 2008; Schuler et al. 2008), but have had limited application for wildlife because of the dearth of toxicity data for wildlife (Awkerman et al. 2008). In some studies, SSDs have been used to derive quality criteria to protect top predators from residues in soils (Jongbloed et al. 1996; Traas et al. 1996). By incorporating interspecies toxicity correlation models, SSDs were developed for wildlife from toxicity data on 23 chemicals (Awkerman et al. 2008). SSDs created for 15 or more wildlife species could give accurate results, whereas data for approximately 7 species can be used to provide only an adequate estimate for some combinations of chemicals and species (Awkerman et al. 2008). In the present study, information for ten species was used to construct the SSD for MeHg, with the log-normal function providing an accepted fit to the distribution, which was tested by use of the Anderson-Darling and Kolmogorov-Smirnov tests. The dietary TRC of 9.56 ng MeHg/g (wwt), derived by SSD, was slightly less than the value of 15.47 ng MeHg/g (wwt) that resulted from

CSA, but the results were similar. Because this assessment, the results of which are reported here, utilized data for ten species, the assessment resolution was no better than 10% (1 of 10). However, an HC_5 was interpolated, which previous work has shown is similar to the threshold value for effects as determined in multispecies tests. This interpolation makes sense, because not all of a compound is environmentally relevant and effects on animals are dependent on the dose and dose rate. Furthermore, animals have the ability to repair damage, and thereby they exhibit some resilience. For these reasons the HC_5 is regarded as being a reasonable surrogate for the adverse effect threshold at the community and ecosystem levels of organization (Giesy et al. 1999).

6 Comparison to Ambient Concentrations in Tissues

To judge the reasonableness of protective guidelines derived by the two methods, and to determine the potential for MeHg to cause adverse effects on fish-eating birds, the derived TRCs were compared to MeHg concentrations measured during monitoring of aquatic environments in China. Concentrations of MeHg in fish tissues, and THg in birds of some Chinese aquatic systems, were collected from the literature. Concentrations of THg have been reported for prey of little egrets in Tai Lake and Poyang Lake, and the geometric means given were 0.24 and 0.10 µg THg/g (dwt), respectively (Zhang et al. 2006). According to a conservative assumption that vertebrates have >50% of the THg as MeHg in muscle (Albers et al. 2007; Eisler 2000), and that the moisture content is 80%, the estimated MeHg concentrations in the prey of little egrets in Tai Lake and Poyang Lake would be 0.024 and 0.01 µg MeHg/g (wwt), respectively. Concentrations of MeHg in fish from Baihua Lake (Yan et al. 2008), Hongfeng Lake (He et al. 2010), Songhua River (Zhu et al. 2012), and Ya-er Lake (Jin et al. 2006) were also available for comparison (Fig. 3). All recorded concentrations of MeHg in fish were greater than the dietary-based TRC value of 9.56 ng MeHg/g (wwt), which was derived by using the SSD. Concentrations of Hg in fish from the Songhua River and Ya-er Lake were significantly greater than this TRC value ($p < 0.05$ by t-test), while concentrations of Hg in fish from Baihua, Hongfeng, and Poyang Lakes approached this TRC value (Fig. 3). Concentrations of MeHg found in fishes of Ya-er Lake (Jin et al. 2006) and prey of little egrets in Tai Lake were greater than the dietary-based TRC of 15.47 ng MeHg/g (wwt) derived by CSA, especially that of Ya-er Lake, which was significantly different from the TRC ($p < 0.05$ by t-test). Therefore, there is a significant risk of MeHg causing adverse effects on avian wildlife populations in the Songhua River, and Ya-er and Tai Lakes, but a lesser risk in Baihua, Hongfeng, and Poyang Lakes. These results are consistent with the pollution characteristics of these water bodies. The Songhua River (Zhu et al. 2012), and Ya-er (Jin et al. 2006) and Tai Lakes are known to be more polluted than Poyang Lake (Zhang et al. 2006). Although both Baihua and Hongfeng Lakes were polluted by Hg from chemical plants in southwest China, the special characteristics of the water environment (e.g., alkaline water, eutrophication, low-dissolved organic carbon, and short food chain) limited accumulation of Hg in fish (He et al. 2010; Yan et al. 2008).

Currently, no data for concentrations of MeHg in blood or feathers of birds in China are available for comparison to the developed criteria. However, studies have been conducted to determine the total concentrations of Hg (THg) in bird feathers, including species located in Poyang, Tai Lakes, the Pearl River Delta (Zhang et al. 2006), Hong Kong and Szechuan (Burger and Gochfeld 1993). Nearly 100% of the Hg that is found in feathers is in the form of MeHg (Herring et al. 2009; Thompson and Furness 1989; Kim et al. 1996). The range of concentrations of THg in feathers was 0.26–4.1 μg THg/g (dwt). The TRC, based on concentrations of THg in feathers that were determined in the present meta-analysis, was expressed as μg THg/g (fresh weight basis). Because information on the feather moisture content of wild birds was not available to convert dry mass to fresh mass, some mallard feathers were collected and dried for 12 h in an oven at 80 °C. The range of moisture content found in mallard feathers was 14–20%. Thus, it was assumed that the moisture content of scapular feather of white ibis was approximately 20%. Given that value, the TRC, based on concentrations expressed on a dry weight basis, would be 3.95 μg THg/g (dwt). Only the concentration of Hg in feathers of little egret from Au Tau in Hong Kong (Connell et al. 2001) exceeded the TRC (Fig. 4). This result is in agreement with a previous risk assessment, in which Hg probably had an adverse effect on the breeding success of the little egret at this site (Connell et al. 2001). Although the concentration of Hg in little egret prey at Tai Lake exceeded both dietary-based TRCs, the level of Hg found in little egret feathers was less than the feather-based TRC. We speculated that this may be due to absorption and metabolism of Hg in the little egret. In conclusion, the TRC values reported in this study can be used as indicators for screening-level risk assessment of avian wildlife in Chinese aquatic systems.

7 Evaluation of Uncertainties

Describing and assessing uncertainty is an important part of deriving and applying TRVs and TRCs (US EPA 1998). Application of qualitative and quantitative expressions of uncertainty compensate for deficiencies in knowledge concerning the accuracy of test results and the data gaps related to the extrapolation of toxicity data among species. If uncertainty factors are identified and are defensible, a more accurate estimate of TRVs and TRCs for protecting wildlife can be achieved. However, when uncertainty factors are applied, they are meant to make criteria more protective without being overly protective; the resulting TRVs and TRCs should be considered to be protective more than predictive of adverse effects under field conditions. According to the US EPA GLI Methodology and GLWQI technical support document for wildlife criteria, three sources of uncertainty were considered for CSA in this study. The first source of uncertainty was associated with interspecies extrapolation. Based on a comparison with other species for which LOAEL and NOAEL values are

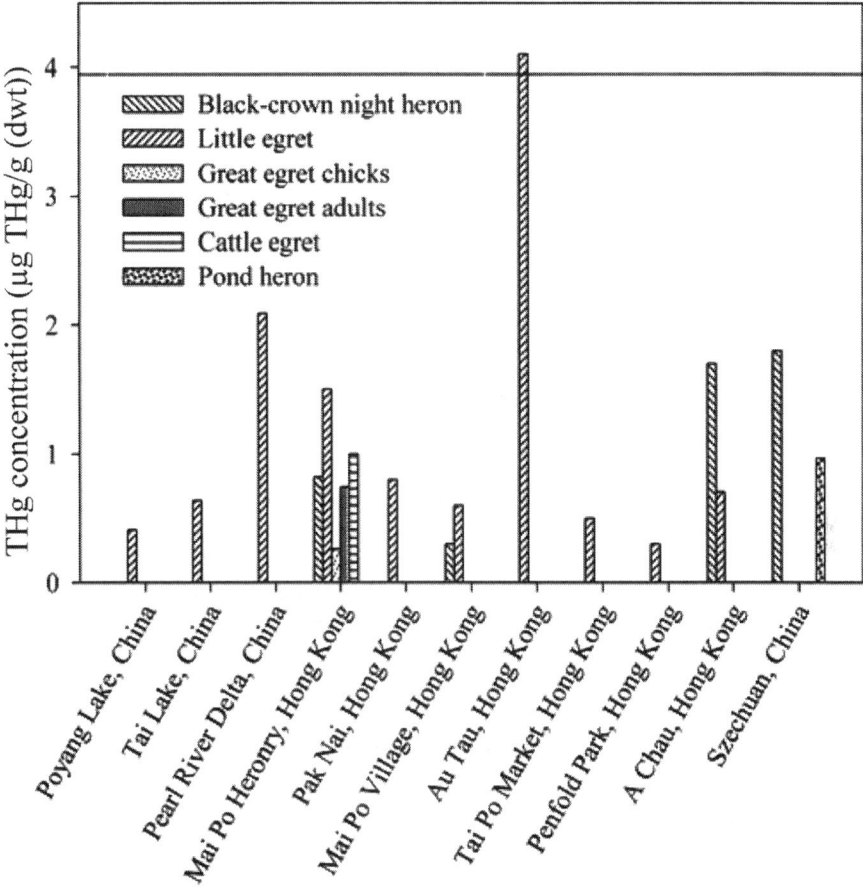

Fig. 4 Total mercury concentrations (THg) found at different sites for bird feathers in China. The *horizontal solid line* represents the toxicity reference criterion (TRC) for feathers. *dwt* dry weight. Poyang, Tai Lakes, and Pearl River Delta (Zhang et al. 2006), Hong Kong and Szechuan (Burger and Gochfeld 1993)

presented (Table 2), the white ibis is the most sensitive species. The results of a study, in which MeHg was injected into eggs, indicates that white ibis is one of the species that are most sensitive to injected MeHg (Heinz et al. 2009). In addition, the results of tests done under laboratory conditions are available from ten species of birds; the white ibis is a piscivorous wading bird that is in the same order as the three representative birds. Therefore, an UF_A of 1.0 was reasonable for interspecies extrapolation of toxicity data to protect other avian species. The second source of uncertainty was associated with LOAEL to NOAEL extrapolation. The LOAEL selected for deriving TRVs was identified from the least exposure dose of 0.05 μg MeHg/g (wwt) in the white ibis study. Compared with the controls, the loss of productivity for this dose of

0.05 μg MeHg/g (wwt) over the 3-year test period was only 13.2% (Frederick and Jayasena 2010). This LOAEL is similar to the NOAEL for common loons exposed to MeHg in the diet (Fig. 1). This LOAEL was similar to the threshold for effects of MeHg on white ibises. In addition, the range of LOAEL to NOAEL ratios for other species was 1.5–6. Thus, a UF_L of 2.0 was used in the extrapolation of the LOAEL to NOAEL. The third source of uncertainty was associated with extrapolation from results of subchronic exposures to chronic exposures. In the white ibis reproductive study, individuals were initially exposed to MeHg when birds were 90 days old, and exposure continued over 3 years, covering three breeding seasons. The value therefore needs no adjustment to cover longer exposure periods. A value of 1.0 was assigned for the UF_S, which could protect the wildlife against chronic effects.

When the SSD approach was used, two sources of uncertainty were explicitly considered, i.e., the relationship between the LOAEL and NOAEL and between subchronic and chronic exposures. Considering the LOAEL to NOAEL ratios of species, a value of 2.0 was assigned to the LOAEL to NOAEL correction factor. A factor of 10 was used to account for the uncertainty in establishing a NOAEL from short-term exposure (<1 month) (Jongbloed et al. 1996). Although the SSD is useful in assessing the range of sensitivities among species during derivation of water quality criteria, much attention has been directed towards extrapolation from laboratory tests to field conditions (Forbes and Forbes 1993; Smith and Cairns 1993). Other uncertainty factors have been used to correct NOAEL data from laboratory tests, and have been used to account for differences in metabolic rate, caloric content of food, and food assimilation efficiency between laboratory and wild species (Traas et al. 1996). Moreover, some studies used a statistical procedure, which is more scientifically defensible, to estimate uncertainty factors so as to obtain precise uncertainty factors and criteria (Calabrese and Baldwin 1994; Dourson and Parker 2007; Gaylor and Kodell 2000). Thus, more research is needed for addressing the suitability of TRVs and TRCs that are derived from laboratory species for protecting avian wildlife in the field.

8 Summary

MeHg is the most biologically available and toxic form of mercury, and has the potential to bioaccumulate and biomagnify as it moves up the food chain. These characteristics result in MeHg exposure to avian wildlife at high trophic levels that can produce adverse effects. The toxicity of MeHg to birds was reviewed, and using available data, TRVs and TRCs were derived for protecting birds in China. The TRV and TRC values were based on concentrations of MeHg in diet (or fish tissue based) and tissues of birds. Two methods were applied to derive TRVs from concentrations in the diet or in tissues. These were the CSA and SSD approaches. Results of published studies show that reproductive productivity of white ibis was the most sensitive endpoint for MeHg exposure, and study results on white ibises were used for deriving the TRV and TRC values, which included applying a UF of 2.0. For the

SSD approach, data for ten species were used to construct the SSD for MeHg, and to calculate the dietary-based TRV and TRC values. Using the CSA approach, the TRV was based on MeHg in the diet and was derived as 5.0 ng MeHg/g (bm)/day; for feathers and blood, the TRVs were 3.16 μg THg/g (wwt), and 0.365 μg THg/g (wwt), respectively. The corresponding TRCs were 15.47 ng MeHg/g (wwt), 3.16 μg THg/g (wwt) and 0.365 μg THg/g (wwt), respectively. The dietary-based TRV and TRC derived by SSD were 3.09 ng MeHg/g (bm)/day and 9.56 ng MeHg/g (wwt), respectively. However, bird tissue residue-based criteria were not available because insufficient MeHg effects data existed to construct an SSD for birds. We compared the criteria derived in our study to those developed by others, and concluded that our results provided more reasonable protection to Chinese avian wildlife. By comparing the criteria values we calculated to actual MeHg levels in fish and bird tissues, we concluded that these criteria values are useful indicators for screening-level risk assessments of avian wildlife in Chinese aquatic systems. The results of this meta-analysis might therefore have important implications for assessing the risk of Hg exposure to birds and for environmental management in China and in other regions. Moreover, because humans and top avian wildlife consumers are at the same trophic level, these criteria may also be used as a reference for human health risk assessment. The diet of birds consists of aquatic species from different trophic levels. However, the structure of the food web for avian wildlife and the environmental factors that affect their exposure to MeHg vary among aquatic systems. Therefore, further research results are needed on the food web structure of avian wildlife in Chinese aquatic systems to provide more insight into what constitutes adequate protection for avian wildlife.

Acknowledgements This work was supported by National Basic Research Program of China (2008CB418200), National Major Scientific Research Program on Environmental Protection Public wolfare C No. 2010090321, and National Natural Science Foundation of China (U0833603, 41130743). Prof. Giesy was supported by the Canada Research Chair program, and at large Chair Professorship at the Department of Biology and Chemistry and State Key Laboratory in Marine Pollution, City University of Hong Kong, and the Einstein Professor Program of the Chinese Academy of Sciences.

References

Adams EM, Frederick PC (2008) Effects of methylmercury and spatial complexity on foraging behavior and foraging efficiency in juvenile white ibises (*Eudocimus albus*). Environ Toxicol Chem 27(8):1708–1712. doi:10.1897/07-466.1

Albers PH, Koterba MT, Rossmann R, Link WA, French JB, Bennett RS, Bauer WC (2007) Effects of methylmercury on reproduction in American kestrels. Environ Toxicol Chem 26(9):1856–1866. doi:10.1897/06-592r.1

Aldenberg T, Jaworska JS (2000) Uncertainty of the hazardous concentration and fraction affected for normal species sensitivity distributions. Ecotoxicol Environ Saf 46(1):1–18. doi:10.1006/eesa.1999.1869

Aldenberg T, Slob W (1993) Confidence limits for hazardous concentrations based on logistically distributed NOEC toxicity data. Ecotoxicol Environ Saf 25(1):48–63

An W, Hu J, Yao F (2006) A method of assessing ecological risk to night heron, *Nycticorax nycticorax*, population persistence from dichlorodiphenyltrichloroethane exposure. Environ Toxicol Chem 25(1):281–286. doi:10.1897/05-043r.1

Awkerman JA, Raimondo S, Barron MG (2008) Development of species sensitivity distributions for wildlife using interspecies toxicity correlation models. Environ Sci Technol 42(9):3447–3452. doi:10.1021/es702861u

Barr JF (1996) Aspects of common loon (*Gavia immer*) feeding biology on its breeding ground. Hydrobiologia 321(2):119–144. doi:10.1007/bf00023169

Barter M, Cao L, Chen L, Lei G (2005) Results of a survey for waterbirds in the lower Yangtze floodplain, China, in January-February 2004. Forktail 21:1–7

Beckvar N, Dillon TM, Read LB (2005) Approaches for linking whole-body fish tissue residues of mercury or DDT to biological effects thresholds. Environ Toxicol Chem 24(8):2094–2105. doi:10.1897/04-284r.1

Blankenship AL, Kay DP, Zwiernik MJ, Holem RR, Newsted JL, Hecker M, Giesy JP (2008) Toxicity reference values for mink exposed to 2,3,7,8-tetrachlodibenzo-p-dioxin (TCDD) equivalents (TEQs). Ecotoxicol Environ Saf 69(3):325–349. doi:10.1016/j.ecoenv.2007.08.017

Bouton SN, Frederick PC, Spalding MG, McGill H (1999) Effects of chronic, low concentrations of dietary methylmercury on the behavior of juvenile great egrets. Environ Toxicol Chem 18(9):1934–1939. doi:10.1002/etc.5620180911

Buekers J, Steen Redeker E, Smolders E (2009) Lead toxicity to wildlife: derivation of a critical blood concentration for wildlife monitoring based on literature data. Sci Total Environ 407(11):3431–3438. doi:10.1016/j.scitotenv.2009.01.044

Burger J, Gochfeld M (1993) Heavy metal and selenium levels in feathers of young egrets and herons from Hong Kong and Szechuan, China. Arch Environ Contam Toxicol 25(3):322–327. doi:10.1007/bf00210724

Burger J, Gochfeld M (1997) Heavy metal and selenium concentrations in feathers of egrets from Bali and Sulawesi, Indonesia. Arch Environ Contam Toxicol 32(2):217–221. doi:10.1007/s002449900178

Calabrese EJ, Baldwin LA (1994) A toxicological basis to derive a generic interspecies uncertainty factor. Environ Health Perspect 102(1):14–17

Caldwell DJ, Mastrocco F, Hutchinson TH, Länge R, Heijerick D, Janssen C, Anderson PD, Sumpter JP (2008) Derivation of an aquatic predicted no-effect concentration for the synthetic hormone, 17α-ethinyl estradiol. Environ Sci Technol 42(19):7046–7054. doi:10.1021/es800633q

CCME (1998) Protocol for derivation of canadian tissue residue guidelines for the protection of wildlife that consume aquatic biota. Canadian Council of Ministers of the Environment, Winnipeg

CCME (2000) Canadian tissue residue guidelines for the protection of wildlife consumers of aquatic biota: Methylmercury. Canadian Council of Ministers of the Environment, Winnipeg

Connell DW, Wong BSF, Lam PKS, Poon KF, Lam MHW, Wu RSS, Richardson BJ, Yen YF (2001) Risk to breeding success of ardeids by contaminants in Hong Kong: evidence from trace metals in feathers. Ecotoxicology 11(1):49–59. doi:10.1023/a:1013745113901

Delnicki D, Reinecke KJ (1986) Mid-winter food use and body weights of mallards and wood ducks in Mississippi. J Wildl Manage 50(1):43–51

Dourson ML, Parker AL (2007) Past and future use of default assumptions and uncertainty factors: default assumptions, misunderstandings, and new concepts. Hum Ecol Risk Assess 13(1):82–87. doi:10.1080/10807030601105480

Dunning JB (1993) CRC handbook of avian body masses. CRC Press I LLC, Boca Raton, FL

Eisler R (ed) (2000) Handbook of chemical risk assessment: Health hazards to humans, plants, and animals, Volume 1: Metals. Lewis Publishers, Boca Raton, FL

Eskeland B, Nafstad I (1978) The modifying effect of multiple generation selection and dietary cadmium on methyl mercury toxicity in Japanese quail. Arch Toxicol 40(4):303–314

Fallacara DM, Halbrook RS, French JB (2011) Toxic effects of dietary methylmercury on immune function and hematology in American kestrels (*Falco sparverius*). Environ Toxicol Chem 30(6):1320–1327. doi:10.1002/etc.494

Feng X (2005) Mercury pollution in China—an overview. In: Pirrone N, Mahaffey KR (eds) Dynamics of mercury pollution on regional and global scales: atmospheric processes and human exposures around the world. Springer US, New York, pp 657–678. doi:10.1007/0-387-24494-8_27

Fimreite N (1970) Effects of methyl mercury treated feed on the mortality and growth of leghorn cockerels. Can J Anim Sci 50(2):387–389. doi:10.4141/cjas70-058

Fimreite N (1971) Effects of dietary methylmercury on ring-necked pheasants, with special reference to reproduction. Canadian Wildlife Service, Ottawa, Occasional paper, no. 9. Department of the Environment

Fimreite N, Karstad L (1971) Effects of dietary methyl mercury on red-tailed hawks. J Wildl Manage 35:293–300

Forbes TL, Forbes VE (1993) A critique of the use of distribution-based extrapolation models in ecotoxicology. Funct Ecol 7(3):249–254

Fournier F, Karasov WH, Kenow KP, Meyer MW, Hines RK (2002) The oral bioavailability and toxicokinetics of methylmercury in common loon (*Gavia immer*) chicks. Comp Biochem Physiol A Mol Integr Physiol 133(3):703–714. doi:10.1016/s1095-6433(02)00140-x

Frederick P, Jayasena N (2010) Altered pairing behaviour and reproductive success in white ibises exposed to environmentally relevant concentrations of methylmercury. Proc R Soc B 278:1851–1857. doi:10.1098/rspb.2010.2189

Frederick P, Campbell A, Jayasena N, Borkhataria R (2011) Survival of white ibises (*Eudocimus albus*) in response to chronic experimental methylmercury exposure. Ecotoxicology 20(2):358–364. doi:10.1007/s10646-010-0586-9

Fujita M (2003) Head bobbing and the body movement of little egrets (*Egretta garzetta*) during walking. J Comp Physiol A 189(1):53–58. doi:10.1007/s00359-002-0376-9

Gaylor DW, Kodell RL (2000) Percentiles of the product of uncertainty factors for establishing probabilistic reference doses. Risk Anal 20(2):245–250. doi:10.1111/0272-4332.202023

Giesy JP, Solomon KR, Coats JR, Dixon KR, Giddings JM, Kenaga EE (1999) Chlorpyrifos: ecological risk assessment in North American aquatic environments. Rev Environ Contam T 160:1–129

Hall LW, Scott MC, Killen WD (1998) Ecological risk assessment of copper and cadmium in surface waters of Chesapeake Bay watershed. Environ Toxicol Chem 17(6):1172–1189. doi:10.1002/etc.5620170626

He T, Wu Y, Pan L, Feng X (2010) Distribution of mercury species and their concentrations in fish in hongfeng reservoir. Journal of Southwest University (Natural Science Edition) 32(7):78–82

Heath JA, Frederick PC, Karasov W (2005) Relationships among mercury concentrations, hormones, and nesting effort of white ibises (*Eudocimus albus*) in the Florida Everglades. The Auk 122(1):255–267

Heinz G (1974) Effects of low dietary levels of methyl mercury on mallard reproduction. Bull Environ Contam Toxicol 11(4):386–392. doi:10.1007/bf01684947

Heinz G (1975) Effects of methylmercury on approach and avoidance behavior of mallard ducklings. Bull Environ Contam Toxicol 13(5):554–564. doi:10.1007/bf01685179

Heinz GH (1976a) Methylmercury: second-generation reproductive and behavioral effects on mallard ducks. J Wildl Manage 40(4):710–715

Heinz GH (1976b) Methylmercury: second-year feeding effects on mallard reproduction and duckling behavior. J Wildl Manage 40:82–90

Heinz GH (1979) Methylmercury: reproductive and behavioral effects on three generations of mallard ducks. J Wildl Manage 43:394–401

Heinz G, Locke LN (1976) Brain lesions in mallard ducklings from parents fed methylmercury. Avian Dis 20(1):9–17

Heinz G, Hoffman D, Klimstra J, Stebbins K, Kondrad S, Erwin C (2009) Species differences in the sensitivity of avian embryos to methylmercury. Arch Environ Contam Toxicol 56(1):129–138. doi:10.1007/s00244-008-9160-3

Heinz G, Hoffman D, Klimstra J, Stebbins K (2010a) Reproduction in mallards exposed to dietary concentrations of methylmercury. Ecotoxicology 19(5):977–982. doi:10.1007/s10646-010-0479-y

Heinz GH, Hoffman DJ, Klimstra JD, Stebbins KR (2010b) Enhanced reproduction in mallards fed a low level of methylmercury: an apparent case of hormesis. Environ Toxicol Chem 29(3):650–653. doi:10.1002/etc.64

Heinz G, Hoffman D, Klimstra J, Stebbins K, Kondrad S, Erwin C (2011) Hormesis associated with a low dose of methylmercury injected into mallard eggs. Arch Environ Contam Toxicol 62(1):141–144. doi:10.1007/s00244-011-9680-0

Herring G, Gawlik DE, Rumbold DG (2009) Feather mercury concentrations and physiological condition of great egret and white ibis nestlings in the Florida Everglades. Sci Total Environ 407(8):2641–2649. doi:10.1016/j.scitotenv.2008.12.043

Hoffman DJ, Spalding MG, Frederick PC (2005) Subchronic effects of methylmercury on plasma and organ biochemistries in great egret nestlings. Environ Toxicol Chem 24(12):3078–3084. doi:10.1897/04-570.1

Jayasena N, Frederick PC, Larkin ILV (2011) Endocrine disruption in white ibises (*Eudocimus albus*) caused by exposure to environmentally relevant levels of methylmercury. Aquat Toxicol 105(3–4):321–327. doi:10.1016/j.aquatox.2011.07.003

Jin L, Liang L, Jiang G, Xu Y (2006) Methylmercury, total mercury and total selenium in four common freshwater fish species from Ya-Er Lake, China. Environ Geochem Health 28(5):401–407. doi:10.1007/s10653-005-9038-5

Jongbloed R, Traas T, Luttik R (1996) A probabilistic model for deriving soil quality criteria based on secondary poisoning of top predators: II. Calculations for dichlorodiphenyltrichloroethane (DDT) and cadmium. Ecotoxicol Environ Saf 34(3):279–306

Kahle S, Becker PH (1999) Bird blood as bioindicator for mercury in the environment. Chemosphere 39(14):2451–2457. doi:10.1016/s0045-6535(99)00154-x

Kannan K, Blankenship AL, Jones PD, Giesy JP (2000) Toxicity reference values for the toxic effects of polychlorinated biphenyls to aquatic mammals. Hum Ecol Risk Assess 6(1):181–201. doi:10.1080/10807030091124491

Kenow KP, Gutreuter S, Hines RK, Meyer MW, Fournier F, Karasov WH (2003) Effects of methyl mercury exposure on the growth of juvenile common loons. Ecotoxicology 12(1):171–181. doi:10.1023/a:1022598525891

Kenow KP, Grasman KA, Hines RK, Meyer MW, Gendron-Fitzpatrick A, Spalding MG, Gray BR (2007a) Effects of methylmercury exposure on the immune function of juvenile common loons (*Gavia immer*). Environ Toxicol Chem 26(7):1460–1469. doi:10.1897/06-442r.1

Kenow KP, Meyer MW, Hines RK, Karasov WH (2007b) Distribution and accumulation of mercury in tissues of captive-reared common loon (*Gavia immer*) chicks. Environ Toxicol Chem 26(5):1047–1055. doi:10.1897/06-193r.1

Kenow KP, Hoffman DJ, Hines RK, Meyer MW, Bickham JW, Matson CW, Stebbins KR, Montagna P, Elfessi A (2008) Effects of methylmercury exposure on glutathione metabolism, oxidative stress, and chromosomal damage in captive-reared common loon (*Gavia immer*) chicks. Environ Pollut 156(3):732–738. doi:10.1016/j.envpol.2008.06.009

Kenow KP, Hines RK, Meyer MW, Suarez SA, Gray BR (2010) Effects of methylmercury exposure on the behavior of captive-reared common loon (*Gavia immer*) chicks. Ecotoxicology 19(5):933–944. doi:10.1007/s10646-010-0475-2

Kim EY, Murakami T, Saeki K, Tatsukawa R (1996) Mercury levels and its chemical form in tissues and organs of seabirds. Arch Environ Contam Toxicol 30(2):259–266. doi:10.1007/bf00215806

Kooijman SALM (1987) A safety factor for LC50 values allowing for differences in sensitivity among species. Water Res 21(3):269–276. doi:10.1016/0043-1354(87)90205-3

Kushlan JA (1977) Population energetics of the American white ibis. Auk 94(1):114–122

Kushlan JA (1978) Feeding ecology of wading birds. In: Sprunt A, Ogden J, Winckler S (eds) Wading birds, vol 7. National Audubon Society Research Report, New York, pp 249–296

Ling C (2004) A conservative, nonparametric estimator for the 5th percentile of the species sensitivity distributions. J Stat Plan Inference 123(2):243–258. doi:10.1016/s0378-3758(03)00148-4

Liu J, Wang D, Sun R (2003) Growth and development of homeothermy in nestlings of Eurasian spoonbill (*Platalea eucorodia*). Zoological Res 24(4):249–253

Liu G, Cai Y, Philippi T, Kalla P, Scheidt D, Richards J, Scinto L, Appleby C (2008) Distribution of total and methylmercury in different ecosystem compartments in the Everglades: implications for mercury bioaccumulation. Environ Pollut 153(2):257–265. doi:10.1016/j.envpol.2007.08.030

Nagy KA (1987) Field metabolic rate and food requirement scaling in mammals and birds. Ecol monogr 57:112–128

Nagy K (2001) Food requirements of wild animals: predictive equations for free-living mammals, reptiles, and birds. Nutr Abstr Rev Ser B 71:21R–31R

Newman MC, Ownby DR, Mézin LCA, Powell DC, Christensen TRL, Lerberg SB, Anderson B-A (2000) Applying species-sensitivity distributions in ecological risk assessment: assumptions of distribution type and sufficient numbers of species. Environ Toxicol Chem 19(2):508–515. doi:10.1002/etc.5620190233

Newsted JL, Jones PD, Coady K, Giesy JP (2005) Avian toxicity reference values for perfluorooctane sulfonate. Environ Sci Technol 39(23):9357–9362. doi:10.1021/es050989v

Posthuma L, Suter GW, Traas TP (2002) Species sensitivity distributions in ecotoxicology. CRC Press LLC, Boca Raton, FL

Rumbold D, Lange T, Axelrad D, Atkeson T (2008) Ecological risk of methylmercury in Everglades National Park, Florida, USA. Ecotoxicology 17(7):632–641. doi:10.1007/s10646-008-0234-9

Sample B, Suter D (1993) Toxicological benchmarks for wildlife. ORNL Oak Ridge National Laboratory (US), Oak Ridge

Sappington KG, Bridges TS, Bradbury SP, Erickson RJ, Hendriks AJ, Lanno RP, Meador JP, Mount DR, Salazar MH, Spry DJ (2011) Application of the tissue residue approach in ecological risk assessment. Integr Environ Assess Manag 7(1):116–140. doi:10.1002/ieam.116

Scheuhammer A (1988) Chronic dietary toxicity of methylmercury in the zebra finch, *Poephila guttata*. Bull Environ Contam Toxicol 40(1):123–130

Scheuhammer AM, Meyer MW, Sandheinrich MB, Murray MW (2007) Effects of environmental methylmercury on the health of wild birds, mammals, and fish. Ambio 36(1):12–19

Schuler L, Hoang T, Rand G (2008) Aquatic risk assessment of copper in freshwater and saltwater ecosystems of South Florida. Ecotoxicology 17(7):642–659. doi:10.1007/s10646-008-0236-7

Scott M (1977) Effects of PCBs, DDT, and mercury compounds in chickens and Japanese quail. Fed Proc 36:1888–1893

Smith EP, Cairns J (1993) Extrapolation methods for setting ecological standards for water quality: statistical and ecological concerns. Ecotoxicology 2(3):203–219. doi:10.1007/bf00116425

Solomon KR, Baker DB, Richards RP, Dixon KR, Klaine SJ, La Point TW, Kendall RJ, Weisskopf CP, Giddings JM, Giesy JP, Hall LW, Williams WM (1996) Ecological risk assessment of atrazine in North American surface waters. Environ Toxicol Chem 15(1):31–76. doi:10.1002/etc.5620150105

Spalding MG, Frederick PC, McGill HC, Bouton SN, McDowell LR (2000a) Methylmercury accumulation in tissues and its effects on growth and appetite in captive great egrets. J Wildl Dis 36(3):411

Spalding MG, Frederick PC, McGill HC, Bouton SN, Richey LJ, Schumacher IM, Blackmore C, Harrison J (2000b) Histologic, neurologic, and immunologic effects of methylmercury in captive great egrets. J Wildl Dis 36(3):423

Spann J, Heath R, Kreitzer J, Locke L (1972) Ethyl mercury p-toluene sulfonanilide: lethal and reproductive effects on pheasants. Science 175(4019):328

Stanton B, de Vries S, Donohoe R, Anderson M, Eichelberger JM (2010) Recommended avian toxicity reference value for cadmium: justification and rationale for use in ecological risk assessments. Hum Ecol Risk Assess 16(6):1261–1277. doi:10.1080/10807039.2010.526499

Stephan CE, Mount DI, Hansen DJ, Gentile JH, Chapman GA, Brungs WA (1985) Guidelines for deriving numerical national water quality criteria for the protection of aquatic organisms and their uses. US EPA, Washington DC

Tan SW, Meiller JC, Mahaffey KR (2009) The endocrine effects of mercury in humans and wildlife. Crit Rev Toxicol 39(3):228–269. doi:10.1080/10408440802233259

Thompson DR, Furness RW (1989) The chemical form of mercury stored in South Atlantic seabirds. Environ Pollut 60(3–4):305–317. doi:10.1016/0269-7491(89)90111-5

Traas T, Luttik R, Jongbloed R (1996) A probabilistic model for deriving soil quality criteria based on secondary poisoning of top predators: I. model description and uncertainty analysis. Ecotoxicol Environ Saf 34(3):264–278

US EPA (1993) Wildlife exposure factors handbook, vol I. US EPA, Office of Research and Development, Washington DC, EPA/600/R-93/187

US EPA (1995a) Final water quality guidance for the great lakes. Fed Regist 60(56): 15366–15425

US EPA (1995b) Great lakes water quality initiative criteria documents for the protection of wildlife, US EPA. Office of Water, Washington DC

US EPA (1995c) Great lakes water quality initiative technical support document for wildlife criteria, US EPA. Office of Water, Washington DC

US EPA (1998) Guidelines for ecological risk assessment, US EPA. Office of Research and Development, Washington DC

US EPA (2003) Attachment 4–5. Ecological Soil Screening Levels (Eco-SSLs) standard operating procedure SOP No.6: Derivation of wildlife toxicity reference value (TRV). OWSER directive 92857-55. US EPA, Washington DC

US EPA (2005) Science advisory board consultation document. proposed revisions to aquatic life guidelines: tissue-based criteria for "bioaccumulative" chemicals. US EPA, Office of Water, Washington DC

Van Vlaardingen PLA, Traas TP, Wintersen AM, Aldenberg T (2004) ETX 2.0. A program to calculate hazardous concentrations and fraction affected, based on normally distributed toxicity Data. National Institute for Public Health and the Environment, Bilthoven, Netherlands

Wagner C, Løkke H (1991) Estimation of ecotoxicological protection levels from NOEC toxicity data. Water Res 25(10):1237–1242. doi:10.1016/0043-1354(91)90062-u

Yan H, Feng X, Liu T, Shang L, Li Z, Li G (2008) Present situation of fish mercury pollution in heavily mercury-contaminated Baihua reservoir in Guizhou. Chin J Ecol 27(8):1357–1361

Yáñez JL, Núñez H, Schlatter RP, Jaksi FM (1980) Diet and weight of American kestrels in central Chile. Auk 97(3):629–631

Zamani-Ahmadmahmoodi R, Esmaili-Sari A, Savabieasfahani M, Bahramifar N (2010) Cattle egret (*Bubulcus ibis*) and little egret (*Egretta garzetta*) as monitors of mercury contamination in Shadegan Wetlands of south-western Iran. Environ Monit Assess 166(1):371–377. doi:10.1007/s10661-009-1008-4

Zhang L, Liu Z (1991) Ecological study on the breeding habits of little egret. J Shanxi Univ 14(2):202–208

Zhang Y, Ruan L, Fasola M, Boncompagni E, Dong Y, Dai N, Gandini C, Orvini E, Ruiz X (2006) Little egrets (*Egretta garzetta*) and trace-metal contamination in wetlands of China. Environ Monit Assess 118(1):355–368. doi:10.1007/s10661-006-1496-4

Zheng W, Kang S, Feng X, Zhang Q, Li C (2010) Mercury speciation and spatial distribution in surface waters of the Yarlung Zangbo River, Tibet. Chin Sci Bull 55(24):2697–2703. doi:10.1007/s11434-010-4001-y

Zhu H, Yan B, Cao H, Wang L (2012) Risk assessment for methylmercury in fish from the Songhua River, China: 30 years after mercury-containing wastewater outfalls were eliminated. Environ Monit Assess 184(1):77–88. doi:10.1007/s10661-011-1948-3

The Biological Effects and Possible Modes of Action of Nanosilver

Carolin Völker, Matthias Oetken, and Jörg Oehlmann

Contents

1 Introduction 81
2 The In Vitro Toxicity of Silver 83
   2.1 Silver as Disruptor of Basal Cell Functions 83
   2.2 The Antibacterial Properties of Silver Compounds 89
3 The In Vivo Toxicity of Silver 92
4 Are Effects Caused by Nanoparticles or Released Silver Ions? 97
5 Conclusions and Future Research 98
6 Summary 99
References 100

1 Introduction

Engineered nanomaterials are increasingly employed in a variety of applications. The size of nanoparticles, by definition, ranges between 1 and 100 nm in at least one dimension (The Royal Society and The Royal Academy of Engineering 2004). Such dimensions result in a high surface area to volume ratio. The subsequent chemical, physical, and biological properties of nanomaterials are unique, and lead to diverse technical applications and prospectively to widespread use in commercial products. In 2004, the production volume of nanomaterials was estimated to be 2,000 t worldwide, and is expected to rise to 58,000 t within the next decade (The Royal Society and The Royal Academy of Engineering 2004).

Currently, the majority of nanotechnology-enabled consumer products are based on nanoscale silver (Woodrow Wilson International Center for Scholars 2011).

C. Völker (✉) • M. Oetken • J. Oehlmann
Department Aquatic Ecotoxicology, Goethe University Frankfurt am Main,
Max-von-Laue-Straße 13, 60438 Frankfurt am Main, Germany
e-mail: c.voelker@bio.uni-frankfurt.de

Because of its antimicrobial properties that were initially used to dress wounds, sanitize medical equipment, and treat water (Gaiser et al. 2009), nanosilver use has been extended to a variety of products, including textiles, cosmetics, food packaging materials, and electronics (Wijnhoven et al. 2009). Silver compounds have long been used for their bactericidal properties (Kim et al. 2002; Silver et al. 2006). The number of applications to which nanosilver is increasingly being put suggests that it has higher activity than its bulk counterpart (Choi et al. 2008). The majority of nanosilver applications require a nanoparticle size range of 1–10 nm; nanosilver in this size range is synthesized by reducing dissolved silver salts (bottom-up technique), usually silver nitrate (Tolaymat et al. 2010). Capping agents are used to prevent silver nanoparticles from aggregating, and the dominant ones are polyvinylpyrrolidone (PVP) and citrate (Tolaymat et al. 2010). These capping agents predominantly lead to a negative particle charge under standard environmental pH conditions (Tolaymat et al. 2010).

As the number and diversity of silver nanoparticle applications increase, concerns about the potential environmental impact of silver nanoparticles are growing (Wijnhoven et al. 2009; Woodrow Wilson International Center for Scholars 2011). Benn and Westerhoff (2008) have demonstrated that silver particles are released from commercially available sock fabrics during the laundering process. Farkas et al. (2011) demonstrated the release of nanosilver from a silver nanowashing machine. Other sources of silver derive from the use and degradation of various consumer products that are disposed of and enter wastewaters (Bradford et al. 2009). Mueller and Nowack (2008) modeled the environmental exposure to engineered nanomaterials. Because estimates for the production and application of nanomaterials vary considerably, the authors' modeling predicted environmental concentrations (PEC) under two exposure scenarios: a realistic one and a high emission scenario. The PEC results for nanosilver were estimated to be in the range from 1.7×10^{-3} to 4.4×10^{-3} µg/m^3 in air, 0.03 to 0.08 µg/L in water, and 0.02 to 0.1 µg/kg in soil.

The environmental behavior and toxicity of silver are known to be influenced by several physicochemical factors of the media in which it exists, including pH, organic carbon content, and cation exchange capacity (Ratte 1999). Toxicity depends on and varies with the type of silver species involved, particularly for free silver ions, which show the highest potential toxicity (Ratte 1999). The ionic form of silver is known to be highly toxic to aquatic organisms (Eisler 1996), and has well-documented antibacterial properties (Bragg and Rainnie 1974; Ghandour et al. 1988; Schreurs and Rosenberg 1982). In 1954, silver was registered as a pesticide in the USA for the first time (U.S. EPA 1993). Recently, an antibacterial product for textile preservation based on nanosilver was conditionally registered. However, the manufacturer was urged to develop further product chemistry, toxicology, exposure, and environmental data as a condition of this registration (U.S. EPA 2010).

The majority of toxicity data available on nanosilver are on bacteria species (Choi et al. 2008; Fabrega et al. 2009; Lok et al. 2006; Morones et al. 2005; Sondi and Salopek-Sondi 2004) and cell lines (Arora et al. 2009; AshaRani et al. 2009a; Bouwmeester et al. 2011; Carlson et al. 2008; Foldbjerg et al. 2011). One proposed mechanism by which nanosilver produces toxicity is by enhancing intracellular levels of reactive oxygen species (ROS). ROS, when formed, produce subsequent cellular damage such as disrupting membrane integrity and damaging proteins and

DNA (Arora et al. 2009; AshaRani et al. 2009a; Bouwmeester et al. 2011; Carlson et al. 2008; Foldbjerg et al. 2011; Gogoi et al. 2006; Lok et al. 2006; Morones et al. 2005; Sondi and Salopek-Sondi 2004).

Aquatic organism toxicity studies have focused primarily on the adverse effects of nanosilver on embryonic development and altered stress enzyme levels in freshwater fish species (Laban et al. 2010; Yeo and Kang 2008). Among freshwater species, *Daphnia* had the highest acute susceptibility to nanosilver, probably because particles are effectively taken up due to their filter-feeding strategy (Griffitt et al. 2008). Unfortunately, limited chronic exposure data are currently available.

In this review, we focus on the biological effects of nanosilver and its possible modes of toxic action. In addition, we identify current knowledge gaps and future research needs. The review is based on peer-reviewed papers found in ISI Web of Science until the end of 2011.

In addition to nanosilver, we also address ionic silver in this review, because of its potential role (silver ions released from nanosilver) in causing toxicity. Furthermore, size dimensions of the nanoparticles used in the cited studies are indicated, because there is evidence that effects of nanoparticles may be size dependent. The size dimensions refer to the actual particle sizes in the test media and size scales are provided as described in the studies. However, when there was no characterization of particle sizes in the test media, primary particle sizes are provided.

2 The In Vitro Toxicity of Silver

2.1 *Silver as Disruptor of Basal Cell Functions*

In vitro test systems allow for specific cellular processes to be directly studied with high reproducibility, and have been used to evaluate the mechanisms by which silver nanoparticles are toxic (Braydich-Stolle et al. 2005; Carlson et al. 2008; Foldbjerg et al. 2011; Hussain et al. 2005). Existing studies have primarily been focused on mammalian cell lines. Because cell types, culture conditions, and types of silver nanoparticles vary, effects noted among studies are not directly comparable. Nevertheless, some conclusions can be drawn.

In vitro study results show that silver nanoparticles are able to enter cells, probably by phagocytosis (AshaRani et al. 2009a; Carlson et al. 2008; Wei et al. 2010) or passive diffusion through the cell membrane (Carlson et al. 2008). Once inside the cell cytoplasm, silver nanoparticles appear in intracellular vesicles and are able to enter organelles like mitochondria and nuclei (Arora et al. 2009; AshaRani et al. 2009a; Carlson et al. 2008; Wei et al. 2010). The entry of silver nanoparticles into cells and their toxic potential is size dependent. Wei et al. (2010) showed that silver microparticles (diameters of 2–20 µm) did not enter mouse fibroblast cells, whereas silver nanoparticles with diameters of 50–100 nm were detected inside the same cells. Studies that have compared different-sized silver nanoparticles have shown smaller particles to be more cytotoxic, probably from the increased reactive surface area of smaller nanoparticles (Carlson et al. 2008; Wei et al. 2010).

Size-dependent toxicity occurred not only for nanosilver but also for nanoparticles of different chemical composition (Griffitt et al. 2008; Xia et al. 2006). However, a particular particle size is not the pivotal factor that determines toxicity, because particles of a similar size that have different chemical composition show different toxicities (Buzea et al. 2007; Griffitt et al. 2008; Liu et al. 2009). Among nanometals, nanosilver is most toxic to bacteria and freshwater organisms, followed by nanoforms of copper and zinc oxide (Griffitt et al. 2008; Kahru and Dubourguier 2010). Titanium dioxide, for example, is less toxic than other nanometals of a similar size. Therefore, the chemical composition and resulting intrinsic properties of nanoparticles greatly influence toxicity (Griffitt et al. 2008).

Silver nanoparticles primarily seem to cause their toxicity by enhancing intracellular levels of reactive oxygen species (ROS). ROS include radicals containing oxygen, such as superoxide ($O_2^{\cdot-}$), hydrogen peroxide (H_2O_2), hydroxyl ($^{\cdot}OH$), or nitroxyl ($NO^{\cdot}$) radicals, and by-products, e.g., alkoxyl radicals ($RO^{\cdot}$) (Lesser 2006; Simon et al. 2000). Radical oxygen intermediates are produced during oxidative metabolism by the reduction of oxygen to water in the mitochondrial respiratory chain (Fridovich 1978). Under normal conditions, nonenzymatic and enzymatic detoxification mechanisms (i.e., reduced glutathione (GSH), superoxide dismutase (SOD), and peroxidases-like catalase) prevent free radical reactions from occurring in cells (Apel and Hirt 2004; Lesser 2006; Simon et al. 2000; Yu 1994). Although free radical reactions may lead to oxidative damage of lipids, proteins, and DNA, ROS may also retain a role in cellular function, and can act as chemical messengers in cellular pathways (Lesser 2006; Simon et al. 2000). ROS are not exclusively generated as metabolic by-products; they also may result from the environmental impact caused by UV radiation or chemicals (Bindhumol et al. 2003; Robertson and Orrenius 2000; Sies 1997). If the amount of ROS exceeds the level of cellular antioxidants, cells will be exposed to oxidative stress.

Several in vitro studies have demonstrated that nanosilver exposure increased intracellular ROS levels (AshaRani et al. 2009b; Carlson et al. 2008; Eom and Choi 2010). Elevated levels of hydrogen peroxide and superoxide, for example, were detected in human fibroblasts after treatment with silver nanoparticles (diameters of 6–20 nm) at concentrations of 25 and 50 µg/mL (AshaRani et al. 2009b). The detection of silver nanoparticles in mitochondria and the resultant effects on mitochondrial function (viz., membrane potential or respiratory chain effects; Arora et al. 2009; Carlson et al. 2008; Foldbjerg et al. 2011; Wei et al. 2010) may also alter ROS production, and thereby cause oxidative stress. Table 1 summarizes studies in which an involvement of ROS in the toxic mode of action of nanosilver was demonstrated.

Other indicators of oxidative stress are increased activities of GSH, GSH-related enzymes, and other enzymes that are involved in cellular antioxidant defense mechanisms. Various mammalian cell-line studies have shown that nanosilver treatment altered the activities of antioxidant enzymes and lipid peroxidation, have increased secretion of inflammatory cytokines/chemokines, and have increased the expression of stress response genes (Arora et al. 2009; AshaRani et al. 2009a; Bouwmeester et al. 2011; Carlson et al. 2008; Foldbjerg et al. 2011). Increased levels of GSH and SOD, for example, were detected in primary mouse fibroblasts and primary mouse

Table 1 Evidence for involvement of ROS (reactive oxygen species) in the toxic mode of action of nanosilver

Cell line	Silver nanoparticles (size)	Evidence for involvement of ROS	References
Human lung fibroblast cells (IMR-90) Human glioblastoma cells (U251)	Starch-capped AgNPs (TEM: 6–20 nm)	Permeability changes of mitochondria leading to disruption of calcium homeostasis probably ROS mediated	AshaRani et al. (2009a)
Human lung fibroblast cells (IMR-90) Human glioblastoma cells (U251)	Starch-capped AgNPs (TEM: 6–20 nm)	Increase in hydrogen peroxide and superoxide production; probably ROS mediated: mitochondrial dysfunction, DNA damage resulting in cell cycle arrest	AshaRani et al. (2009b)
Human lung carcinoma epithelial-like cell line (A549)	PVP-coated AgNPs (stock solution (MilliQ), TEM: 69 ± 3 nm, DLS: 121 ± 6 nm; RPMI 1,640 media, DLS: 149 ± 37 nm)	Increased levels of ROS; cytotoxicity, bulky DNA adducts and ROS levels reduced by pretreatment with antioxidant NAC	Foldbjerg et al. (2011)
Jurkat T cells	AgNPs (DLS: 28–35 nm)	Increased levels of ROS leading to increased levels of NF-κB, Nrf-2 which activate MAPK causing apoptosis	Eom and Choi (2010)
Primary mouse fibroblasts Primary mouse liver cells	AgNPs (DLS: 6.5–43.8 nm, average size of 16.6 nm)	Increased levels of GSH, SOD	Arora et al. (2009)
Mouse fibroblast cells (L929)	AgNPs (TEM: 50–100 nm)	DNA damage (possibly ROS mediated) leading to cell cycle arrest and apoptosis	Wei et al. (2010)
Mouse lymphoma cell line (L5178Y thymidine kinase (tk)$^{+/-}$-3.7.2C cells) Human bronchial epithelial cells (BEAS-2B)	AgNPs (manufacturer: <100 nm)	DNA damage (possibly ROS mediated) and cytotoxicity	Kim et al. (2010)
Rat alveolar macrophages	Hydrocarbon-coated AgNPs (SEM: primary sizes 15, 30, 55 nm, larger agglomerates in suspension)	Increased levels of ROS and decreased levels of GSH in cells exposed to AgNPs (15 nm)	Carlson et al. (2008)
Baby hamster kidney cells (BHK21) Human colon adenocarcinoma cells (HT29)	AgNPs (TEM: 10–15 nm)	Apoptosis with involvement of caspases could be ROS mediated	Gopinath et al. (2008)
Danio rerio embryos (whole organism)	AgNPs, supporter material TiO$_2$ (TEM: 10–20 nm)	Increased catalase activity indicates oxidative stress-related toxicity	Yeo and Kang (2008)
Caenorhabditis elegans (whole organism)	AgNPs (uncoated, DLS: 14~20 nm)	Increased sod-3 gene expression indicates oxidative stress-related toxicity	Roh et al. (2009)

liver cells, respectively, after spherical silver nanoparticle (6.5–43.8 nm, average size 16.6 nm) treatment (Arora et al. 2009). In contrast, Carlson et al. (2008) correlated the increased ROS levels in rat alveolar macrophages with depleted GSH levels, after treatment with hydrocarbon-coated spherical-silver nanoparticles with a primary size of 15 nm. As the authors suggested, this could be the result of silver nanoparticles reacting with GSH-maintenance enzymes, for example, direct binding to GSH-reductases. In conclusion, increased ROS levels that are accompanied by depleted GSH levels produce oxidative stress.

Foldbjerg et al. (2011) observed minor cytotoxic effects of nanosilver (149±37 nm) on the human lung carcinoma epithelial cell line (A549) after adding N-acetylcysteine (NAC), a precursor for GSH; this suggests that the observed cytotoxicity was ROS mediated. The authors suggested direct binding of silver to thiol groups of cysteine as another possible reason for the appearance of minor toxic effects after NAC treatment. Exposure to NAC also inhibited bulky DNA adducts that emerged after nanosilver treatment of A549 cells, suggesting that this effect was also ROS mediated.

In several in vitro studies performed with nanosilver, a strong correlation between enhanced ROS levels and apoptosis was detected (Carlson et al. 2008; Foldbjerg et al. 2011; Hsin et al. 2008; Sanpui et al. 2011). Enhanced levels of nuclear factor-kappaB (NF-κB) and nuclear factor-E2-related factor-2 (Nrf-2) were detected in Jurkat T cells that showed increased ROS levels after nanosilver (28–35 nm) treatment. These transcriptional factors are known to activate mitogen-activated protein kinase (MAPK), which is relevant to the induction of apoptosis (Ashkenazi and Dixit 1998; Eom and Choi 2010; Janssen-Heininger et al. 2000). Eom and Choi (2010) revealed an elevated level of p38 MAPK in nanosilver-treated cells that showed apoptosis. Additionally, ROS-induced DNA damage may have produced cell cycle arrest, and may have contributed to the observed apoptosis in nanosilver-treated cells.

Apoptosis induced by DNA damage, with resulting cell cycle arrest, also occurred in mouse fibroblasts (L929) treated with 100 μg/mL nanosilver (diameters of 50–100 nm) (Wei et al. 2010). Nanosilver-induced DNA damage and cytotoxicity also occurred in mouse lymphoma cells (L5178Y) and in human bronchial epithelial cells (BEAS-2B) (Kim et al. 2010). Furthermore, nanosilver-treated cells displayed decreased mitochondrial function (Arora et al. 2009). Disruption of the mitochondrial respiratory chain by silver nanoparticles was suggested by Wei et al. (2010) and Carlson et al. (2008). Such disruption may enhance ROS levels and produce cellular damage, but could also be the cause of a mitochondrial-driven apoptosis from cytochrome C release and caspase cascade activation. The role of mitochondria in nanosilver-mediated apoptosis was also evaluated by AshaRani et al. (2009a). Nanosilver treatment of human lung fibroblasts (IMR-90) and human glioblastoma cells (U251) disrupted calcium homeostasis, probably from mitochondrial permeability changes caused by oxidative stress. Calcium transients in mitochondria could impair mitochondrial function, leading to higher ROS levels and inhibiting ATP synthesis (Orrenius et al. 1992). Furthermore, mitochondrial membrane permeability from calcium overload led to the release of apoptogenic factors like cytochrome C, which initiates the activation of caspases (Belizário et al. 2007).

The general role of ROS in causing apoptosis has been evaluated in several previous studies (Buttke and Sandstrom 1994; Fadeel et al. 1998; Sakon et al. 2003; Simon et al. 2000). ROS are responsible for activating caspase cascades (Fadeel et al. 1998; Simon et al. 2000), which have a critical role in causing programmed cell death (Cohen 1997). Apoptosis is known to result from moderate oxidative stress, whereas severe oxidative stress produces necrosis (Bonfoco et al. 1995; Curtin et al. 2002). Necrosis is associated with inflammation and is characterized by cell swelling and lysis, whereas apoptosis is an active cellular process that leads to morphological cell changes (viz., cell shrinkage, membrane blebbing, nuclear condensation, DNA fragmentation), and to the formation of apoptotic bodies that are engulfed by phagocytic cells (Robertson and Orrenius 2000).

The type (apoptosis vs. necrosis) of nanosilver-induced cell death appears to be dependent on the particle concentration with which cell lines are treated. Nanosilver-treated (spherical, 6.5–43.8 nm, average size 16.6 nm) primary mouse fibroblasts and primary mouse liver cells showed a dose-dependent form of cell death (Arora et al. 2009). Concentrations of 3.12 µg/mL (primary fibroblasts) and 12.5 µg/mL (primary liver cells) produced an apoptotic cell population; the necrotic concentration, at which total lack of caspase-3 activity occurred, was much higher (100 µg/mL in primary fibroblasts, 500 µg/mL in primary liver cells). Apoptotic bodies, cell membrane blebbing, and condensed chromatin occurred in baby hamster kidney (BHK21) and human colon adenocarcinoma (HT29) cell lines after treatment with 11.0 µg/mL silver nanoparticles (10–15 nm) (Gopinath et al. 2008). Furthermore, caspase gene expression increased in nanosilver-treated cells. Necrotic cells resulted from treatment with higher concentrations (>44.0 µg/mL).

In summary, the results of in vitro cytotoxicity studies indicate that one toxic mechanism of nanosilver may be a dose-dependent programmed cell death driven by several apoptotic pathways that are potentially induced by ROS as summarized in Table 1 and Fig. 1. Similar results were obtained for nickel ferrite nanoparticles and zinc oxide nanoparticles in in vitro studies (Ahamed et al. 2010; Xia et al. 2008). As observed for nanosilver, zinc oxide nanoparticles and zinc ions accumulated in cell organelles and produced oxidative stress, mitochondrial damage, and enhanced calcium release (Xia et al. 2008). Titanium dioxide nanoparticles are redox active, capable of generating ROS (Farré et al. 2009), and also become internalized into cells and organelles (Xia et al. 2008). However, the in vitro and in vivo adverse effects elicited by titanium dioxide are relatively minor, and are only observed at high concentrations (Heinlaan et al. 2008; Ivask et al. 2010; Xia et al. 2008).

In contrast to other nanometals that cause increased ROS levels, cerium oxide nanoparticles appear to protect cells from oxidative stress by suppressing ROS generation (Xia et al. 2008).

The antioxidant activity of cerium oxide is explained by its mixed valence state (trivalent (3+) and tetravalent (4+)), and its ability to change its oxidation state (Tarnuzzer et al. 2005). Additionally, cerium oxide with a higher Ce^{3+}/Ce^{4+} ratio shows superoxide dismutase mimetic capabilities, and higher efficiency than the authentic enzyme (Korsvik et al. 2007). Although the adverse effects caused by increased ROS levels were observed for all nanometals (except cerium oxide), the

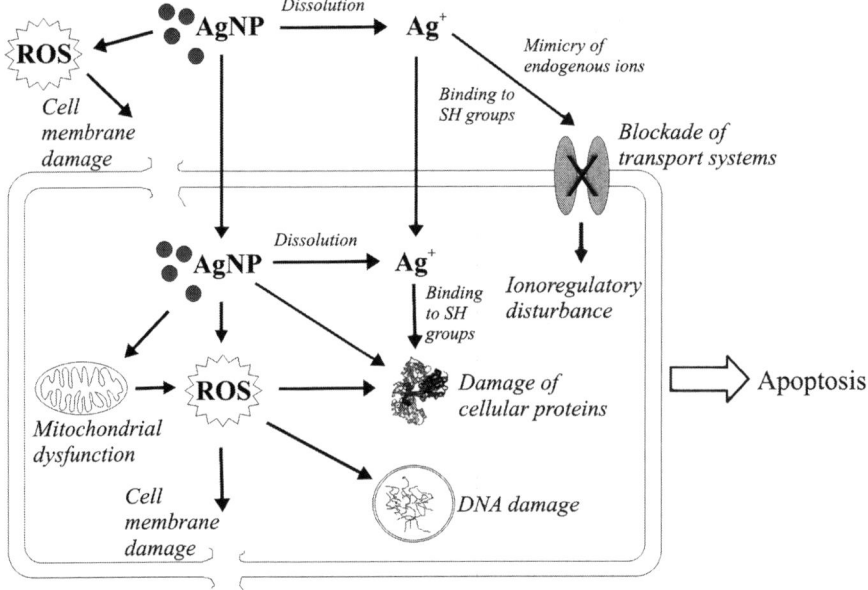

Fig. 1 Possible modes of action of nanosilver toxicity

toxicities differed significantly and were dependent on the particles' chemical composition.

The release of metal ions from nanoparticle surfaces is considered to be a further cause that influences toxicity. Indeed, higher acute toxicities are observed for nanomaterials of higher solubility (viz., zinc oxide, copper oxide), not excluding a possible involvement of both, released ions and the particulate form (Auffan et al. 2009; Brunner et al. 2006).

In general, ionic as well as nanoparticulate forms of heavy metals are capable of producing ROS and subsequent oxidative damage to cellular structures, and of inducing stress proteins and activating protein kinases (Stohs and Bagchi 1995). Another important mechanism of metal toxicity at the cellular level is the replacement or mimicry of essential ions (e.g., lead's well-known mimicry of calcium (Clarkson 1993)). We also note that metals unfold their toxic potential by an incomplete mimicry of endogenous ions in some metabolic steps that leads to, for example, a blockade of a particular metabolic process (Clarkson 1993). Hussain et al. (1994) considered ionic silver mimicry, and the replacement of endogenous ions such as K^+ and Na^+, as one reason for the inhibition of isolated Na^+/K^+-ATPase, after they treated the enzyme with silver nitrate. Because binding of the silver to the enzyme is reversible by adding cysteine, silver ions may block the transport system by binding to the enzyme's thiol groups (Hussain et al. 1994). It is well documented that silver ions show a strong affinity to free thiol groups, since these compounds neutralize the biological effects of silver ions (Hussain et al. 1992; Liau et al. 1997).

Moutin et al. (1989) also hypothesized that ionic silver has the property of structurally mimicking endogenous ions. These authors investigated the effects of silver ions on sarcoplasmic reticulum vesicles that were prepared from rabbit skeletal muscle. They found a concentration-dependent ability of silver ions to trigger calcium release, suggesting that silver ions act on the same site as do calcium ions. This conclusion was confirmed by inhibition of calcium release at higher silver concentrations, which led to a result similar to that shown for calcium-induced calcium release. However, disruption of calcium homeostasis could also result from impairment of calcium translocases (located in membranes) by ROS, as reported for other heavy metals (Viarengo and Nicotera 1991). ROS are capable of oxidizing the SH groups of Ca^{2+}/Mg^{2+}-ATPases, leading to dysfunction and altered calcium levels. In addition, silver ions could bind to the thiol groups of the enzyme, and thereby disrupt its function.

Moreover, it has been documented that silver ions form complexes with isolated DNA (Jensen and Davidson 1966; Yamane and Davidson 1962). These complexes are chemically and biologically reversible (Jensen and Davidson 1966), and there is an indication that the helical structure of the DNA is not disrupted (Yamane and Davidson 1962). In contrast to other cations, Ag-DNA complexes are not formed by an interaction with silver and phosphate groups. Rather, silver binds specifically to purine and pyrimidine bases, probably by replacing the hydrogen bond between complementary base pairs (Jensen and Davidson 1966; Luk et al. 1975; Yamane and Davidson 1962).

It is not yet clear to what extent the toxic mode of action of nanosilver is comparable to that of ionic silver, or whether it is related to the release of silver ions. Studies that compared the effects of silver in its ionic forms and nanoforms show contradictory results, and this will be addressed later in this chapter. However, results of in vitro studies reveal that nanosilver and silver ions, respectively, affect cellular components and functions. Whether these in vitro results can be extrapolated to live animal studies is not yet clear.

To date, only two in vivo studies have been performed that provide evidence for oxidative stress as an important mechanism for nanosilver toxicity: Yeo and Kang (2008) exposed zebrafish embryos (*Danio rerio*) to nanosilver (10–20 nm) and observed an increase of catalase activity, indicating an involvement of ROS in the toxicity of nanosilver. Comparable results are reported by Roh et al. (2009), using the soil nematode *Caenorhabditis elegans* as model organism. After exposure to nanosilver (0.1 and 0.5 mg/L, 14–20 nm) *C. elegans* showed an increased expression of the superoxide dismutases-3 (sod-3) gene. Increased sod-3 gene expression was linked to the involvement of oxidative stress since superoxide dismutases are known to scavenge ROS (Lesser 2006).

2.2 *The Antibacterial Properties of Silver Compounds*

The antibacterial properties of silver compounds are well documented: silver has long been used to treat wounds and disinfect water (Kim et al. 2002; Silver et al. 2006). The antibacterial action of silver probably derives from silver ions attaching to the negatively charged bacterial cell wall, where they can disrupt its permeability

(Ratte 1999). In addition, silver ions are presumed to enter bacterial cells by an essential copper transport system (Ghandour et al. 1988). The toxic mechanism in bacteria is by inhibition of the respiratory chain, collapse of the proton motive force, and interference with phosphate uptake (Bragg and Rainnie 1974; Ghandour et al. 1988; Schreurs and Rosenberg 1982). Because of the high affinity that silver ions have for thiol groups (Liau et al. 1997), the interference with respiratory enzymes could be mediated by silver ions binding to thiol sites on these enzymes (Holt and Bard 2005; Kim et al. 2008b). This was confirmed by a study with *Escherichia coli*, wherein effects of the silver ions on phosphate uptake and exchange were reversed by the addition of thiols (Schreurs and Rosenberg 1982). However, study results indicate that silver does not exclusively act with thiols (Schreurs and Rosenberg 1982).

Feng et al. (2000) documented effects on DNA molecules of *Staphylococcus aureus* and *E. coli* after treatment with silver ions, and showed that the affected DNA molecules condensed and lost their replicating ability. In addition, proteins that surrounded the nuclear region were expressed, perhaps to protect the DNA from the silver ions. In general, effects of the silver ions were more profound on Gram-negative *E. coli* cells, indicating a better protection of Gram-positive *S. aureus* against silver penetration (Feng et al. 2000). This conclusion is in agreement with the results of a study from Jung et al. (2008). Gram-negative cells only possess a thin peptidoglycan layer between their inner and outer membranes, whereas Gram-positive cells possess a thick peptidoglycan layer and lack an outer membrane (Li et al. 2010b; Thiel et al. 2007). In Gram-negative bacteria the outer membrane serves to protect from agents that would normally damage the peptidoglycan layer (Li et al. 2010b). However, in the case of silver ions, the size of the peptidoglycan layer seems to be of particular importance in protecting against the penetration of silver (Feng et al. 2000).

Considering the well-documented bactericidal effects of silver ions, the numerous studies that have dealt with silver nanoparticles have been focused on bacteria. Such studies have either had the goal of evaluating nanoparticle effectiveness for antibacterial applications (Jain and Pradeep 2005; Lee et al. 2007), or of elucidating the environmental impact of the particles, e.g., effects on nitrifying organisms in wastewater treatment plants (Choi et al. 2008). Additional information on the antibacterial effects of silver nanoparticles is presented below. Moreover, a comprehensive look at this topic is provided by Marambio-Jones and Hoek (2010), who reviewed the antibacterial effects of silver nanomaterials in detail.

The interaction of silver nanoparticles with bacterial cell walls has been documented in several studies. Silver particles are known to have attached to the cell wall of Gram-negative bacteria (*E. coli*) (Choi et al. 2008; Gogoi et al. 2006), which resulted in the formation of pits and cell death (Choi et al. 2008; Gogoi et al. 2006; Sondi and Salopek-Sondi 2004). Such adhesions and interactions with bacterial cell surfaces are possibly due to electrostatic forces that exist between the negatively charged cell surface of the bacteria and the silver nanoparticles (Thiel et al. 2007). Furthermore, silver nanoparticles were found to bind to and accumulate in bacterial cell membranes (Morones et al. 2005; Sondi and Salopek-Sondi 2004). When this occurred, effects were produced on membrane structure, permeability, and leakage of cellular components like reducing sugars and proteins (Li et al. 2010b; Morones et al. 2005; Sondi and Salopek-Sondi 2004). Nanosilver may inactivate respiratory chain dehydrogenase (Li et al. 2010b), which results in inhibition of respiration and growth. In addition, nanosilver could affect phosphate lipids of the membrane and deactivate membrane enzymes, finally leading to cell death (Li et al. 2010b).

According to the mechanism hypothesized for silver ions, the interaction of nanosilver with components of bacterial membranes is suggested to be mediated by binding to thiol groups. Bacterial membranes are rich in sulfur-containing proteins, and may be preferential sites for silver nanoparticles (Morones et al. 2005). In general, thiol groups show metal complexing properties (Deratani and Sebille 1981; Jiménez et al. 1997), and may therefore contribute to the antibacterial effects observed to occur for other nanometals. Both zinc and copper oxide nanoparticles show high toxicities to bacteria, followed by aluminum and nickel oxide (Baek and An 2011; Heinlaan et al. 2008; Jiang et al. 2009). Zinc oxide particles were found to attach to bacterial surfaces, although at a lower intensity than aluminum oxide nanoparticles (Baek and An 2011). The attachment of nanoparticles to bacterial surfaces may have toxic manifestations, but is only one aspect of their toxic potential. As evidence for this, the more soluble zinc oxide particles showed lower attachment, but higher toxicity than did aluminum oxide particles (Baek and An 2011).

Studying the effects of nanosilver (diameter of 9.3 ± 2.8 nm) on bacterial membranes (*E. coli*), Lok et al. (2006) found that nanosilver induced the collapse of proton motive force, which decreased cellular potassium and ATP levels. Intact *E. coli* cells normally maintain their membrane potential by a high content of intracellular potassium. Membrane destabilization and the absence of ATP observed after nanosilver treatment resulted in the accumulation of envelope protein precursors in the cytoplasm. Normally, after conversion to mature forms, newly synthesized envelope proteins are incorporated into the outer membrane of bacteria. The conversion to mature forms and the translocation across the cytoplasm membrane requires a membrane potential and energy in form of ATP (Zimmermann and Wickner 1983). Nanosilver treatment disrupted this process, and produced an accumulation of protein precursors (Lok et al. 2006). The detrimental effects of nanosilver on cell membranes may also occur in nitrifying bacteria (Choi and Hu 2008). Interference with bacterial membranes may also apply for nanoparticles of different chemical composition; altered membrane permeability of bacterial membranes was reported for zinc oxide (Huang et al. 2008; Liu et al. 2009).

In addition to interacting with bacterial cell walls and membranes, silver nanoparticles are able to enter bacterial cells (Lok et al. 2006; Morones et al. 2005; Shrivastava et al. 2007; Sondi and Salopek-Sondi 2004). In mammalian cell lines, the uptake rate of silver nanoparticles is size dependent. Morones et al. (2005) revealed that silver particles having an average size of 1–10 nm bound to the membranes of *E. coli*, *Pseudomonas aeruginosa*, *Vibrio cholera*, and *Salmonella typhus*, and these particles were incorporated into the cells. As nanoparticle size decreased, the toxicity increased; this relationship was linked to the greater reactive surface area of the smaller particles (Lok et al. 2007; Morones et al. 2005). Pal et al. (2007) estimated a 109-fold increased surface area as the size of a spherical particle was reduced from

10 µm to 10 nm, which was concomitantly followed by enhanced antibacterial activity. Size-dependent results were also obtained in a study in which *E. coli* and *S. aureus* were treated with silver nanoparticles of three diameters, viz., 7, 29, and 89 nm (Martínez-Castañón et al. 2008). Silver nanoparticles having a size of 7 nm showed the lowest minimum inhibition concentrations (MIC) (viz., 6.25 and 7.5 µg/mL) for *E. coli* and *S. aureus*, respectively. MIC increased with particle size.

As documented for ionic silver, Gram-negative bacteria show higher susceptibilities to nanosilver than do Gram-positive bacteria. When treated with 89-nm-sized particles, *S. aureus* (Gram-positive) showed a MIC of 33.7 µg/mL; values that were approximately threefold those of Gram-negative *E. coli* cells (11.8 µg/mL) (Martínez-Castañón et al. 2008). Shrivastava et al. (2007) demonstrated strong inhibition of Gram-negative *E. coli* cells at 25 µg/mL of nanosilver, whereas Gram-positive *S. aureus* only showed partial inhibition at a concentration of 100 µg/mL.

The mechanisms by which silver nanoparticles unfold their toxic potential inside bacterial cells may also involve interference with sulfur-containing proteins, as has been observed to occur for ionic silver (Morones et al. 2005). This suggests DNA damage, and possible consequences could be disturbance of cell division and cell death (Morones et al. 2005). In addition, research suggests that ROS may be involved in the toxicity of silver nanoparticles to nitrifying bacteria cultures (Choi and Hu 2008), *E. coli* (Hwang et al. 2008), and *P. aeruginosa* (Kora and Arunachalam 2011). Similar observations were made for *E. coli* cells exposed to copper and zinc oxide nanoparticles (Ivask et al. 2010).

In general, size, specific surface area, and shape are important factors that influence the silver nanoparticle toxicity to bacteria (Fabrega et al. 2009; Pal et al. 2007). Differently shaped silver nanoparticles (e.g., spherical, rod shaped, and truncated triangular) inhibited *E. coli* differentially (Pal et al. 2007). Truncated triangular particles showed the strongest antibacterial activity, with complete inhibition of *E. coli* at a concentration of 10 µg/mL. In contrast, spherical particles only reduced growth at a level of 125 µg/mL. Rod-shaped particles did not fully inhibit *E. coli* cells, even at concentrations of 1,000 µg/mL. These differences in the activity of differently shaped particles may be explained by the different atomic structure that is characteristic of different shaped particles.

In summary, the toxic effects produced by nanosilver are primarily from the interference nanosilver has with bacterial cell membranes, and susceptibilities of the bacteria differ as different membrane structures (Gram-negative vs. Gram-positive) are encountered.

3 The In Vivo Toxicity of Silver

Ionic silver is one of the most toxic metals to aquatic organisms (Eisler 1996). Studies with silver nitrate show acute effective concentrations in the low microgram-per-liter range (Bury et al. 1999; Davies et al. 1978; Morgan et al. 1997; Nebeker et al. 1983; Zhao and Wang 2011). The most acutely sensitive freshwater organisms are cladocerans and amphipods (Bianchini et al. 2002; Ratte 1999). The severity of the toxic effects depends on the amount of free silver ions present, which is influenced by the physicochemical parameters of the surrounding medium (Brauner and Wood 2002; Bury et al. 1999; Eisler 1996). When testing the toxicity of silver nitrate, adding food reduces the toxic effects in the test systems (Bianchini and Wood 2002; Hook and Fisher 2001). The reason for the reduction is thought to derive from the presence of organic material that alters silver bioavailability, and enhances the complexation of silver ions by algae.

Mechanisms by which toxic effects are mediated were evaluated in freshwater fish, and one important mechanism by which silver causes toxicity appears to be inhibition of branchial enzymes that are involved in ion transport (Brauner and Wood 2002; Morgan et al. 1997). Normally, sodium and chloride ions are mainly transported across the gills by means of branchial sodium-potassium pump (Na^+/K^+-ATPase), which is directly related to the uptake of these ions (Bianchini and Wood 2003). Silver ions at concentrations of 2 and 10 µg/L, respectively, inhibited gill enzymes like Na^+/K^+-ATPase and carbonic anhydrase in rainbow trout (*Oncorhynchus mykiss*), leading to inhibition of active Na^+ and Cl^- uptake (Morgan et al. 1997). The silver-mediated inhibition of Na^+/K^+-ATPase was also demonstrated by Hussain et al. (1994), who studied the isolated enzyme in vitro. When gill system enzymes were disrupted, ion uptake from water was inhibited in vivo, and resulted in a net loss of ions and death of the organisms (Bianchini and Wood 2003).

An ionoregulatory disturbance induced by silver occurred not only in fish, but in daphnids as well; the daphnids showed a decrease in whole-body sodium concentration after treatment with 5 µg silver nitrate/L (Bianchini and Wood 2002). As the authors suggested, silver and sodium ions may share the same mechanism of transport across the gills, resulting in a quick accumulation of silver in *Daphnia magna* during the experiment. The activity of whole body Na^+/K^+-ATPase increased by 60% from the loss of the sodium ions, probably as a compensatory response to the reduced sodium level. The ionoregulatory disturbance in daphnids is probably mediated by silver mimicking sodium in Na^+ channels that inhibits the function of these channels.

As observed for silver ions, nanoparticulate forms of silver also show high toxic effects on freshwater species. The high toxicity to such aquatic species is in contrast to what occurs in mammals, where the in vivo toxicity of nanosilver is relatively low. For example, a 28-day inhalation study performed with Sprague-Dawley rats that were exposed to concentrations of nanosilver (average size 60 nm) up to 1,000 mg/kg did not show any changes in body weight or hematology and blood biochemical values (Kim et al. 2008a). The only observation was a gender-related difference in accumulation of silver in the kidneys. Female rats showed a twofold higher accumulation compared to male rats, but it was not elucidated as to whether this accumulation was animal-size dependent or if hormones were involved.

The concentrations that are effective in producing silver nanoparticle effects on aquatic organisms vary and depend on the physicochemical properties of the particular nanoparticles used in the tests (Allen et al. 2010; Gaiser et al. 2009). How silver nanoparticles are prepared and how the animals are dosed also affects the intensity of the observed effects (Roh et al. 2009). Effect values range from the low microgram-per-liter range (toxicity comparable to silver nitrate; Allen et al. 2010; Li et al. 2010a) to the high microgram-per-liter range (Gaiser et al. 2011; Zhao and Wang 2011). The effective concentrations obtained for different types of silver nanoparticle dispersions on aquatic organisms are summarized in Table 2.

Allen et al. (2010) studied the effects of different nanosilver dispersions either prepared from uncoated particles or those with different surface coatings. The authors found that the aggregation state of the particles that was affected by particle size produced different LC_{50} values for *D. magna*. Smaller particles had a higher reactive surface and were probably more bioavailable. Laban et al. (2010) also showed differential LC_{50} values for fathead minnow (*Pimephales promelas*) embryos that were affected by how test solutions were prepared. Sonicated nanosilver dispersions led to tenfold higher toxicity (LC_{50} 1.25–1.36 mg/L) than did stirred solutions (LC_{50} 9.40–10.6 mg/L). Stirred solutions showed higher aggregation rates of the nanoparticles. Since aggregation of nanoparticles is positively correlated with their tendency to settle out of the water in which they are dispersed (Chen and Elimelech 2006; Keller et al. 2010), the organisms may have been exposed to lower nanosilver concentrations. Sonicated samples produced to a more stable colloidal suspension and, therefore, a higher particle concentration was delivered to the organisms (Laban et al. 2010). Gaiser et al. (2011) showed size-dependent effects for both nanosilver particles that had a primary size of 35 nm (agglomerated in the test medium to 588 nm) and for microsilver-sized particles of 811 nm. The smaller agglomerates showed approximately a tenfold higher toxicity and increased mortality levels at concentrations of 0.1 mg/L.

Li et al. (2010a) did not find a size-dependent effect on *D. magna* exposed to particles having primary sizes of 36, 52, and 66 nm, but they obtained a relatively low median LC_{50} value (~3 μg/L). The absence of size-dependent effects is explained by aggregation of the particles to diameters of 438, 378, and 553 nm, after 24 h. The low LC_{50} value may have resulted from an extremely fast silver nanoparticle uptake by *D. magna*, as has been shown to occur with silver ions (in <1 h) in previous studies (Glover and Wood 2005). The filter-feeding strategy of daphnids renders particle uptake from the water phase very effective, and, therefore, may explain why daphnids are the most susceptible of organisms to nanosilver exposure (Griffitt et al. 2008).

In most studies, LC_{50} values for daphnids were in the low microgram-per-liter-range (Table 2) (Allen et al. 2010; Griffitt et al. 2008; Li et al. 2010a). Bivalve molluscs are also a filter-feeding taxonomic group, and similarly show a high susceptibility to nanosilver exposure as a result of their feeding strategy. Ringwood et al. (2010) demonstrated adverse effects from nanosilver exposure (diameter of 25 nm, stabilized with sodium citrate). The effects noted were on embryonic development and lysosomal integrity of adult hepatopancreas tissues in oysters (*Crassostrea virginica*) at concentrations of 1.6 μg/L and 0.16 μg/L, respectively. In contrast, fish (viz., zebrafish (*D. rerio*) and fathead minnow (*P. promelas*)) showed tenfold to 100-fold higher lethal concentrations (Griffitt et al. 2008; Laban et al. 2010).

Sublethal effects, derived from silver nanoparticles, and their possible mechanisms were studied in zebrafish. *D. rerio* embryos showed lower hatch rates and

Table 2 Effective concentrations of nanosilver observed in vivo

Test organism	Silver nanoparticles (size)	Effective concentration	References
Caenorhabditis elegans	AgNPs (uncoated, DLS: 14~20 nm)	Significant decrease in reproduction at 0.05; 0.1; 0.5 mg/L	Roh et al. (2009)
Crassostrea virginica (embryos and adults)	AgNPs (stabilized with sodium citrate, DLS: 25 nm)	Significant effects on embryonic development at 1.6 µg/L Lysosomal destabilization (adult oysters, hepatopancreas cells) at 0.16 µg/L	Ringwood et al. (2010)
Daphnia magna neonates	Citrate-coated AgNPs (DLS: 5.94–39.75 nm) Coffee-coated AgNPs 1:100 dilution (DLS: 101.5–773.6 nm) Uncoated AgNPs (Sigma-Aldrich, DLS: 681.4–5,412) Coated AgNPs (Sigma-Aldrich, coating unknown, DLS: 39.39–249.8 nm)	LC_{50} (48 h) 1.1 µg/L LC_{50} (48 h) 1.0 µg/L LC_{50} (48 h) 1.4 µg/L (filtered 100 nm); 16.7 µg/L (unfiltered) LC_{50} (48 h) 4.4 µg/L (filtered 100 nm); 31.5 µg/L (unfiltered)	Allen et al. (2010)
Daphnia magna neonates Cyprius carpio	AgNPs (uncoated, nominal size 35 nm), micro-Ag (nominal size 0.6–1.6 µm)	Daphnia magna (96 h): 60% mortality at 0.1 mg/L (AgNPs); 80% mortality 1 mg/L (micro Ag) Cyprius carpio showed Ag (both forms) in liver, intestine, gills, and gall bladder	Gaiser et al. (2009)
Daphnia magna neonates	AgNPs (uncoated, nominal size 35 nm, DLS: 588 nm in reconstituted hard water), micro-Ag (nominal size 0.6–1.6 µm, DLS: 811 nm in reconstituted hard water)	96 h acute tests AgNPs: 10 and 1 mg/L 100% mortality, 0.1 mg/L 56.7% mortality 96 h acute tests micro Ag: 10 mg/L 100% mortality, 1 mg/L 80% mortality, 0.1 mg/L no significant toxicity	Gaiser et al. (2011)
Daphnia magna (7-day adults)	Carbonate-coated AgNPs (TEM: 20 nm)	Uptake rate constant (k_u): 0.060 L/g/h at 2, 10, and 40 µg/L; 2.2 L/g/h at 160 and 500 µg/L >70% of AgNP in daphnids was accumulated through the dietary route	Zhao and Wang (2010)

(continued)

Table 2 (continued)

Test organism	Silver nanoparticles (size)	Effective concentration	References
Daphnia magna (neonates)	Carbonate-coated AgNPs (TEM: 20 nm, DLS: 40–50 nm)	Significant decrease in body length and reproduction at 50 µg/L	Zhao and Wang (2011)
Daphnia pulex (adults)	AgNPs (coated with metal oxide, primary size 20–30 nm, major particle diameters observed in suspension (SEM) 44.5, 216, 94.5 nm)	*Daphnia pulex*: LC_{50} (48 h) 0.040 mg/L	Griffitt et al. (2008)
Ceriodaphnia dubia (neonates)		*Ceriodaphnia dubia*: LC_{50} (48 h) 0.067 mg/L	
Danio rerio (adult and juvenile <24 h)		*Danio rerio* (48 h): LC_{50} (adult) 7.07 mg/L, LC_{50} (juvenile) 7.20 mg/L	
Danio rerio (embryos)	Starch-coated AgNPs, BSA-coated AgNPs (TEM: both 5–20 nm)	LC_{50} (dependent on growth stage): 25–50 mg/L	AshaRani et al. (2008)
Danio rerio (embryos)	AgNPs, supporter material TiO_2 (TEM: 10–20 nm)	Reduced hatch rates and larval abnormalities at 10 ng/L	Yeo and Kang (2008)
Pimephales promelas (embryos)	AgNPs (uncoated, TEM: 29–100 nm, majority 31–50 nm) AgNPs (uncoated, TEM: 21–280 nm, majority 21–60 nm)	LC_{50} (96 h): 9.4 (stirred), 1.25 (sonicated) mg/L LC_{50} (96 h): 10.6 (stirred), 1.36 (sonicated) mg/L	Laban et al. (2010)

several larval abnormalities after exposure to nanosilver (10–20 nm) (Yeo and Kang 2008). Roh et al. (2009) used the soil nematode *C. elegans* as a model organism to study nanosilver toxicity. After exposure to nanosilver (0.1 and 0.5 mg/L, diameter 14–20 nm), *C. elegans* showed decreased reproductive ability, accompanied by increased expression of the superoxide dismutases-3 (sod-3) gene. Laban et al. (2010) demonstrated a concentration-dependent increase of larval abnormalities in nanosilver-exposed embryos of *P. promelas*. In addition, uptake of particles ranging in size from 29 to 100 nm to 21 to 280 nm was observed. Possible uptake mechanisms suggested by the authors were diffusion across membrane pores, or active uptake by endocytosis. Entry of silver nanoparticles by diffusion into cells, or by endocytosis, was proposed by AshaRani et al. (2008), after they evaluated nanosilver-treated *D. rerio* embryos. The nanosilver-treated embryos displayed phenotypic defects like abnormal body axes, degeneration of body parts, pericardial edema, and cardiac arrhythmias. In addition, an increase in apoptosis and necrosis in the body parts that accumulated blood was detected. The particles having an average size of 5–20 nm were detected in the brain, heart, yolk, and blood of embryos, and a high particle accumulation occurred in the nucleus. The authors opined that DNA damage and chromosomal aberrations, derived from the nanoparticles located in the nucleus, may have accounted for the observed toxic effects.

Uptake of silver nanoparticles by *D. magna* was studied by Zhao and Wang (2010), and these authors used their results to develop hypotheses on possible mechanisms.

Low concentrations of nanosilver (2–40 µg/L) in the water phase led to uptake rates in proportion to the nanosilver concentration, whereas uptake rates at higher concentrations (160 and 500 µg/L) increased disproportionately. Besides endocytosis as possible route of nanosilver uptake the authors proposed direct ingestion into the gut as explanation for the higher uptake rates at higher concentrations. The authors also investigated the dietary ingestion of nanosilver via algal food. Because nanosilver showed a strong accumulation in or on algae in the experiments, the authors concluded that nanosilver incorporated into the daphnids mainly via the diet. Ingested silver nanoparticles could not be completely depurated and led to sublethal effects like decreased reproduction and growth in *D. magna* treated with 5 and 50 µg/L (Zhao and Wang 2011). In addition, low depuration of nanosilver from the daphnids may be important in the potential transport of nanosilver along the aquatic food chain (Zhao and Wang 2010, 2011).

In summary, there are only a few studies in which the toxic mode of action of nanosilver was evaluated as an in vivo exposure. Of those studies that were completed results indicated that ROS are involved in producing the observed toxicity in the test systems.

4 Are Effects Caused by Nanoparticles or Released Silver Ions?

Release of silver ions from nanoparticle surfaces is considered to be important to nanosilver toxicity. In several studies that have dealt with silver nanoparticles, the degree to which silver ions were released was regarded to play a role in toxicity (AshaRani et al. 2008; Bouwmeester et al. 2011; Carlson et al. 2008; Eom and Choi 2010; Foldbjerg et al. 2011; Griffitt et al. 2008; Laban et al. 2010; Morones et al. 2005; Roh et al. 2009). Treating mammalian cell lines in vitro with silver nitrate often revealed effects that were comparable to those obtained for nanosilver (Bouwmeester et al. 2011; Carlson et al. 2008; Eom and Choi 2010). Notwithstanding the similarity of action, there also appears to be a unique nanoparticle effect. Eom and Choi (2010) and Foldbjerg et al. (2011) demonstrated that silver nanoparticles induced higher titers of ROS than did silver ions. In addition, increased expressions of transcriptional factors like Nrf-2 and NF-κB, and accumulating DNA damage that produced apoptosis, were observed for nanosilver, but were not evident effects of silver ions (Eom and Choi 2010).

Results with bacteria also suggest that toxic effects may not exclusively result from the release of silver ions from nanoparticle surfaces (Choi and Hu 2008; Fabrega et al. 2009; Lok et al. 2007; Morones et al. 2005). Choi and Hu (2008) revealed higher toxic effects from silver nanoparticles than silver ions on nitrifying bacteria. The authors hypothesized that the small size (<5 nm) of the uncharged nanoparticles potentially enhanced effective uptake, whereas charged ions are not easily transported across the cell membrane. In contrast, Hwang et al. (2008) found an equal induction of ROS, when they compared the effects of silver nanoparticles with those of silver ions released from silver nitrate. This result was confirmed by Pal et al. (2007), who proposed that silver ions constitute the cause of the detrimental effects.

The results of in vivo studies indicate that silver nanoparticles and silver ions have different mechanisms of toxicity. AshaRani et al. (2008) did not find that silver ions were involved in the toxicity of nanosilver to *D. rerio*, because none of the larval abnormalities found for nanosilver could be shown to occur for silver nitrate. Reproductive effects on *C. elegans* were slightly more prominent, when the organism was treated with silver nanoparticles than with silver ions provided as silver nitrate (Roh et al. 2009). Stress-related gene expression also differed, suggesting different toxicity mechanisms. Different uptake mechanisms for silver nanoparticles and silver ions were proposed by Zhao and Wang (2010). They showed that nanoparticle uptake was about 4.3-fold lower than silver ion uptake. Griffitt et al. (2008) suggested that the toxicity of silver nanoparticles did not result from silver ion release in several aquatic species. The ion concentrations released from silver nanoparticles during the test were lower than those of silver nitrate that produced lethal effects. Laban et al. (2010) hypothesized that ions from silver nitrate caused greater toxicity than did ions released from silver nanoparticles. The toxicity of silver ions from silver nitrate was three times higher to *P. promelas* as compared to the silver ions from nanoparticles. This effect is explained by the more rapid dissociation of silver nitrate than silver ions from silver nanoparticles.

From the foregoing, we conclude that the observed toxic effects of silver nanoparticles are caused by both silver ions and the particulate form. The release of free silver ions may contribute to some, but not to all, toxic effects observed for nanosilver (Gaiser et al. 2011; Laban et al. 2010). This conclusion may also apply to other nanometals, as similar observations were made for zinc oxide nanoparticles, nanonickel, and nanocopper (Franklin et al. 2007; Griffitt et al. 2008).

5 Conclusions and Future Research

Based on the results of a comprehensive literature study on toxic effects of nanosilver and silver ions, we offer the following conclusions as to what the several mechanisms of action are for silver nanoparticles (Fig. 1); some may also apply for other nanometals:

1. *ROS generation.* Increased ROS levels caused by nanosilver may account for observed cases of cellular damage and apoptosis. ROS generation and oxidative stress may either result from the catalytic properties of silver nanoparticles, an effect of mitochondrial dysfunction caused by nanosilver, or constitute a combination of both mechanisms.
2. *Interaction with cellular enzymes.* The evidence suggests that silver nanoparticles may interact with cellular enzymes. Since silver ions show strong affinities to free thiol groups, nanosilver may show similar effects. Binding to thiol groups can lead to damage and inactivation of proteins and enzymes and, therefore, to

cellular damage. In addition, silver ions bind to DNA molecules, specifically to purine and pyrimidine bases.
3. *Mimicry of endogenous ions.* Researchers have also demonstrated that silver is probably capable of structurally mimicking endogenous ions, like calcium, sodium, or potassium ions. Such action can block transport systems and induce ionoregulatory disturbances.
4. *Release of silver ions.* When silver ions are released from nanosilver forms, organisms may be affected by the released ions. Silver ions may affect organisms by similar modes of action as reported for nanosilver. However, some evidence indicates that the effects produced may not totally derive from the released silver ions.

Finally, the results reported in this review are primarily from in vitro test systems. Such systems have limitations and do not allow one to confidently predict probable effects on whole organisms in vivo. Based on the reviewed studies, we believe there is compelling evidence that nanosilver is active via an ROS-mediated toxic mechanism, in both in vivo and in vitro systems. Of course, other possible mechanisms may be operating and may be involved either separately or in tandem.

To evaluate whether an increased production and application of silver nanoparticles pose an environmental risk, additional in vitro data are needed. Most importantly, observations are needed in in vivo test systems to better predict and prevent future adverse effects on communities and ecosystems as nanosilver use increases. Special emphasis should be given to chronic in vivo studies, ideally covering the entire life cycle of test organisms. Furthermore, systematic studies investigating dietary uptake of silver nanoparticles and the potential biomagnification risk in the food web should be undertaken.

6 Summary

Novel physicochemical and biological properties have led to a versatile spectrum of applications for nanosized silver particles. Silver nanoparticles are applied primarily for their antimicrobial effects, and a variety of commercially available products have emerged. To better predict and prevent possible environmental impacts from silver nanoparticles that are derived from increasing production volumes and environmental release, more data on the biological effects are needed on appropriate model organisms.

We examined the literature that addressed the adverse effects of silver nanoparticles on different levels of biological integration, including in vitro and in vivo test systems. Results of in vitro studies indicate a dose-dependent programmed cell death induced by oxidative stress as main possible pathway of toxicity. Furthermore, silver nanoparticles may affect cellular enzymes by interference with free thiol groups and mimicry of endogenous ions.

Similar mechanisms may apply for antibacterial effects produced by nanosilver. These effects are primarily from the interference nanosilver has with bacterial cell membranes.

Few in vivo studies have been performed to evaluate the toxic mode of action of nanosilver or to provide evidence for oxidative stress as an important mechanism of nanosilver toxicity. Organisms that are most acutely sensitive to nanosilver toxicity are the freshwater filter-feeding organisms.

Both in vitro and in vivo studies have demonstrated that silver ions released from nanoparticle surfaces contribute to the toxicity of nanosilver. Contradictory results exist on the extent to which silver ions contribute to toxicity, and, indeed, some findings indicate a unique nanoparticle effect.

For an adequate evaluation of the environmental impact of nanosilver, greater emphasis should be placed on combining mechanistic investigations that are performed in vitro, with results obtained in in vivo test systems. Future in vivo test system studies should emphasize long-term exposure scenarios. Moreover, the dietary uptake of silver nanoparticles and the potential to bioaccumulate through the food web should be examined in detail.

Acknowledgements The authors wish to thank D. Whitacre and H. Hollert for critically commenting on and editing the manuscript. This work was financially supported by the German National Academic Foundation.

References

Ahamed M, AlSalhi MS, Siddiqui MKJ (2010) Silver nanoparticle applications and human health. Clin Chim Acta 411:1841–1848

Allen HJ, Impellitteri CA, Macke DA, Heckman JL, Poynton HC, Lazorchak JM, Govindaswamy S, Roose DL, Nadagouda MN (2010) Effects from filtration, capping agents, and presence/absence of food on the toxicity of silver nanoparticles to *Daphnia magna*. Environ Toxicol Chem 29:2742–2750

Apel K, Hirt H (2004) Reactive oxygen species: metabolism, oxidative stress, and signal transduction. Annu Rev Plant Biol 55:373–399

Arora S, Jain J, Rajwade JM, Paknikar KM (2009) Interactions of silver nanoparticles with primary mouse fibroblasts and liver cells. Toxicol Appl Pharm 236:310–318

AshaRani PV, Wu YL, Gong Z, Valiyaveettil S (2008) Toxicity of silver nanoparticles in zebrafish models. Nanotechnology 19(25):255102

AshaRani PV, Hande MP, Valiyaveettil S (2009a) Anti-proliferative activity of silver nanoparticles. BMC Cell Biol 10:65

AshaRani PV, Mun GLK, Hande MP, Valiyaveettil S (2009b) Cytotoxicity and genotoxicity of silver nanoparticles in human cells. ACS Nano 3:279–290

Ashkenazi A, Dixit VM (1998) Death receptors: signaling and modulation. Science 281:1305–1308

Auffan M, Rose J, Wiesner MR, Bottero JY (2009) Chemical stability of metallic nanoparticles: a parameter controlling their potential cellular toxicity *in vitro*. Environ Pollut 157:1127–1133

Baek YW, An YJ (2011) Microbial toxicity of metal oxide nanoparticles (CuO, NiO, ZnO, and Sb_2O_3) to *Escherichia coli*, *Bacillus subtilis*, and *Streptococcus aureus*. Sci Total Environ 409:1603–1608

Belizário JE, Alves J, Occhiucci JM, Garay-Malpartida M, Sesso A (2007) A mechanistic view of mitochondrial death decision pores. Braz J Med Biol Res 40:1011–1024

Benn TM, Westerhoff P (2008) Nanoparticle silver released into water from commercially available sock fabrics. Environ Sci Technol 42:4133–4139

Bianchini A, Wood CM (2002) Physiological effects of chronic silver exposure in *Daphnia magna*. Comp Biochem Phys C 133:137–145

Bianchini A, Wood CM (2003) Mechanism of acute silver toxicity in *Daphnia magna*. Environ Toxicol Chem 22:1361–1367

Bianchini A, Bowles KC, Brauner CJ, Gorsuch JW, Kramer JR, Wood CM (2002) Evaluation of the effect of reactive sulfide on the acute toxicity of silver (I) to *Daphnia magna*. Part II: toxicity results. Environ Toxicol Chem 21:1294–1300

Bindhumol V, Chitra KC, Mathur PP (2003) Bisphenol A induces reactive oxygen species generation in the liver of male rats. Toxicology 188:117–124

Bonfoco E, Krainc D, Ankarcrona M, Nicotera P, Lipton SA (1995) Apoptosis and necrosis: two distinct events induced, respectively, by mild and intense insults with N-methyl-D-aspartate or nitric oxide/superoxide in cortical cell cultures. Proc Natl Acad Sci U S A 92:7162–7166

Bouwmeester H, Poortman J, Peters RJ, Wijma E, Kramer E, Makama S, Puspitaninganindita K, Marvin HJP, Peijnenburg Ad ACM, Hendriksen PJM (2011) Characterization of translocation of silver nanoparticles and effects on whole-genome gene expression using an *in vitro* intestinal epithelium coculture model. ACS Nano 5:4091–4103

Bradford A, Handy RD, Readman JW, Atfield A, Mühling M (2009) Impact of silver nanoparticle contamination on the genetic diversity of natural bacterial assemblages in estuarine sediments. Environ Sci Technol 43:4530–4536

Bragg PD, Rainnie DJ (1974) The effect of silver ions on respiratory-chain of *Escherichia coli*. Can J Microbiol 20:883–889

Brauner CJ, Wood CM (2002) Effect of long-term silver exposure on survival and ionoregulatory development in rainbow trout (*Oncorhynchus mykiss*) embryos and larvae, in the presence and absence of added dissolved organic matter. Comp Biochem Phys C 133:161–173

Braydich-Stolle L, Hussain S, Schlager JJ, Hofmann MC (2005) *In vitro* cytotoxicity of nanoparticles in mammalian germline stem cells. Toxicol Sci 88:412–419

Brunner TJ, Wick P, Manser P, Spohn P, Grass RN, Limbach LK, Bruinink A, Stark WJ (2006) *In vitro* cytotoxicity of oxide nanoparticles: comparison to asbestos, silica, and the effect of particle solubility. Environ Sci Technol 40:4374–4381

Bury NR, Galvez F, Wood CM (1999) Effects of chloride, calcium, and dissolved organic carbon on silver toxicity: comparison between rainbow trout and fathead minnows. Environ Toxicol Chem 18:56–62

Buttke TM, Sandstrom PA (1994) Oxidative stress as a mediator of apoptosis. Immunol Today 15:7–10

Buzea C, Pacheco II, Robbie K (2007) Nanomaterials and nanoparticles: sources and toxicity. Biointerphases 2(4):MR17–MR71

Carlson C, Hussain SM, Schrand AM, Braydich-Stolle LK, Hess KL, Jones RL, Schlager JJ (2008) Unique cellular interaction of silver nanoparticles: size-dependent generation of reactive oxygen species. J Phys Chem B 112:13608–13619

Chen KL, Elimelech M (2006) Aggregation and deposition kinetics of fullerene (C_{60}) nanoparticles. Langmuir 22:10994–11001

Choi O, Hu ZQ (2008) Size dependent and reactive oxygen species related nanosilver toxicity to nitrifying bacteria. Environ Sci Technol 42:4583–4588

Choi O, Deng KK, Kim NJ, Ross L, Surampalli RY, Hu ZQ (2008) The inhibitory effects of silver nanoparticles, silver ions, and silver chloride colloids on microbial growth. Water Res 42:3066–3074

Clarkson TW (1993) Molecular and ionic mimicry of toxic metals. Annu Rev Pharmacol 33:545–571

Cohen GM (1997) Caspases: the executioners of apoptosis. Biochemical J 326:1–16

Curtin JF, Donovan M, Cotter TG (2002) Regulation and measurement of oxidative stress in apoptosis. J Immunol Methods 265:49–72

Davies PH, Goettl JP, Sinley JR (1978) Toxicity of silver to rainbow trout (*Salmo gairdneri*). Water Res 12:113–117

Deratani A, Sebille B (1981) Metal ion extraction with a thiol hydrophilic resin. Anal Chem 53:1742–1746

Eisler R (1996) Silver hazards to fish, wildlife, and invertebrates: a synoptic review. Contaminant Hazard Reviews. Biological Report 32. National Biological Service, Washington, DC, pp 1–44

U.S. Environmental Protection Agency (1993) Reregistration eligibility decision fact sheet. Silver. Office of Prevention, Pesticides And Toxic Substances. EPA-738-F-93-005. http://www.epa.gov/oppsrrd1/REDs/factsheets/4082fact.pdf. Acessed 26 Sep 2012

U.S. Environmental Protection Agency (2010) Pesticide news story: EPA proposes conditional registration of nanosilver pesticide product. http://epa.gov/oppfead1/cb/csb_page/updates/2010/nanosilver.html. Accessed 20 Mar 2012

Eom HJ, Choi J (2010) P38 MAPK activation, DNA damage, cell cycle arrest and apoptosis as mechanisms of toxicity of silver nanoparticles in Jurkat T cells. Environ Sci Technol 44:8337–8342

Fabrega J, Fawcett SR, Renshwa JC, Lead JR (2009) Silver nanoparticle impact on bacterial growth: effect of pH, concentration, and organic matter. Environ Sci Technol 43:7285–7290

Fadeel B, Åhlin A, Henter JI, Orrenius S, Hampton MB (1998) Involvement of caspases in neutrophil apoptosis: regulation by reactive oxygen species. Blood 92:4808–4818

Farkas J, Peter H, Christian P, Urrea JAG, Hassellöv M, Tuoriniemi J, Gustafsson S, Olsson E, Hylland K, Thomas KV (2011) Characterization of the effluent from a nanosilver producing washing machine. Environ Int 37:1057–1062

Farré M, Gajda-Schrantz K, Kantiani L, Barceló D (2009) Ecotoxicity and analysis of nanomaterials in the aquatic environment. Anal Bioanal Chem 393:81–95

Feng QL, Wu J, Cheng GQ, Cui FZ, Kim TN, Kim JO (2000) A mechanistic study of the antibacterial effect of silver ions on *Escherichia coli* and *Staphylococcus aureus*. J Biomed Mater Res 52:662–668

Foldbjerg R, Dang DA, Autrup H (2011) Cytotoxicity and genotoxicity of silver nanoparticles in the human lung cancer cell line, A549. Arch Toxicol 85:743–750

Franklin NM, Rogers NJ, Apte SC, Batley GE, Gadd GE, Casey PS (2007) Comparative toxicity of nanoparticulate ZnO, bulk ZnO, and $ZnCl_2$ to a freshwater microalga (*Pseudokirchneriella subcapitata*): the importance of particle solubility. Environ Sci Technol 41:8484–8490

Fridovich I (1978) The biology of oxygen radicals. Science 201:875–880

Gaiser BK, Fernandes TF, Jepson M, Lead JR, Tyler CR, Stone V (2009) Assessing exposure, uptake and toxicity of silver and cerium dioxide nanoparticles from contaminated environments. Environ Health 8:S2

Gaiser BK, Biswas A, Rosenkranz P, Jepson MA, Lead JR, Stone V, Tyler CR, Fernandes TF (2011) Effects of silver and cerium dioxide micro- and nano-sized particles on *Daphnia magna*. J Environ Monitor 13:1227–1235

Ghandour W, Hubbard JA, Deistung J, Hughes MN, Poole RK (1988) The uptake of silver ions by *Escherichia coli* K12: toxic effects and interaction with copper ions. Appl Microbiol Biot 28:559–565

Glover CN, Wood CM (2005) Accumulation and elimination of silver in *Daphnia magna* and the effect of natural organic matter. Aquat Toxicol 73:406–417

Gogoi SK, Gopinath P, Paul A, Ramesh A, Ghosh SS, Chattopadhyay A (2006) Green fluorescent protein-expressing *Escherichia coli* as a model system for investigating the antimicrobial activities of silver nanoparticles. Langmuir 22:9322–9328

Gopinath P, Gogoi SK, Chattopadhyay A, Ghosh SS (2008) Implications of silver nanoparticle induced cell apoptosis for *in vitro* gene therapy. Nanotechnology 19:075104

Griffitt RJ, Luo J, Gao J, Bonzongo JC, Barber DS (2008) Effects of particle composition and species on toxicity of metallic nanomaterials in aquatic organisms. Environ Toxicol Chem 27:1972–1978

Heinlaan M, Ivask A, Blinova I, Dubourguier HC, Kahru A (2008) Toxicity of nanosized and bulk ZnO, CuO and TiO_2 to bacteria *Vibrio fischeri* and crustaceans *Daphnia magna* and *Thamnocephalus platyurus*. Chemosphere 71:1308–1316

Holt KB, Bard AJ (2005) Interaction of silver(I) ions with the respiratory chain of *Escherichia coli*: an electrochemical and scanning electrochemical microscopy study of the antimicrobial mechanism of micromolar Ag^+. Biochemistry 44:13214–13223

Hook SE, Fisher NS (2001) Sublethal effects of silver in zooplankton: importance of exposure pathways and implications for toxicity testing. Environ Toxicol Chem 20:568–574

Hsin YH, Chen CF, Huang S, Shih TS, Lai PS, Chueh PJ (2008) The apoptotic effect of nanosilver is mediated by a ROS- and JNK-dependent mechanism involving the mitochondrial pathway in NIH3T3 cells. Toxicol Lett 179:130–139

Huang ZB, Zheng X, Yan DH, Yin GF, Liao XM, Kang YQ, Yao YD, Huang D, Hao BQ (2008) Toxicological effect of ZnO nanoparticles based on bacteria. Langmuir 24:4140–4144

Hussain S, Anner RM, Anner BM (1992) Cysteine protects Na, K-ATPase and isolated human lymphocytes from silver toxicity. Biochem Bioph Res Co 189:1444–1449

Hussain S, Meneghini E, Moosmayer M, Lacotte D, Anner BM (1994) Potent and reversible interaction of silver with pure Na, K-ATPase and Na, K-ATPase-liposomes. Biochim Biophys Acta 1190:402–408

Hussain SM, Hess KL, Gearhart JM, Geiss KT, Schlager JJ (2005) *In vitro* toxicity of nanoparticles in BRL 3A rat liver cells. Toxicol in Vitro 19:975–983

Hwang ET, Lee JH, Chae YJ, Kim YS, Kim BC, Sang BI, Gu MB (2008) Analysis of the toxic mode of action of silver nanoparticles using stress-specific bioluminescent bacteria. Small 4:746–750

Ivask A, Bondarenko O, Jepihhina N, Kahru A (2010) Profiling of the reactive oxygen species-related ecotoxicity of CuO, ZnO, TiO_2, silver and fullerene nanoparticles using a set of recombinant luminescent *Escherichia coli* strains: differentiating the impact of particles and solubilised metals. Anal Bioanal Chem 398:701–716

Jain P, Pradeep T (2005) Potential of silver nanoparticle-coated polyurethane foam as an antibacterial water filter. Biotechnol Bioeng 90:59–63

Janssen-Heininger YMW, Poynter ME, Baeuerle PA (2000) Recent advances towards understanding redox mechanisms in the activation of nuclear factor κB. Free Radical Bio Med 28:1317–1327

Jensen RH, Davidson N (1966) Spectrophotometric, potentiometric, and density gradient ultracentrifugation studies of the binding of silver ion by DNA. Biopolymers 4:17–32

Jiang W, Mashayekhi H, Xing BS (2009) Bacterial toxicity comparison between nano- and microscaled oxide particles. Environ Pollut 157:1619–1625

Jiménez I, Gotteland M, Zarzuelo A, Uauy R, Speisky H (1997) Loss of the metal binding properties of metallothionein induced by hydrogen peroxide and free radicals. Toxicology 120:37–46

Jung WK, Koo HC, Kim KW, Shin S, Kim SH, Park YH (2008) Antibacterial activity and mechanism of action of the silver ion in *Staphylococcus aureus* and *Escherichia coli*. Appl Environ Microb 74:2171–2178

Kahru A, Dubourguier HC (2010) From ecotoxicology to nanoecotoxicology. Toxicology 269:105–119

Keller AA, Wang HT, Zhou DX, Lenihan HS, Cherr G, Cardinale BJ, Miller R, Ji ZX (2010) Stability and aggregation of metal oxide nanoparticles in natural aqueous matrices. Environ Sci Technol 44:1962–1967

Kim BR, Anderson JE, Mueller SA, Gaines WA, Kendall AM (2002) Literature review—efficacy of various disinfectants against *Legionella* in water systems. Water Res 36:4433–4444

Kim YS, Kim JS, Cho HS, Rha DS, Kim JM, Park JD, Choi BS, Lim R, Chang HK, Chung YH, Kwon IH, Jeong J, Han BS, Yu IJ (2008a) Twenty-eight-day oral toxicity, genotoxicity, and gender-related tissue distribution of silver nanoparticles in Sprague-Dawley rats. Inhal Toxicol 20:575–583

Kim JY, Lee C, Cho M, Yoon J (2008b) Enhanced inactivation of *E. coli* and MS-2 phage by silver ions combined with UV-A and visible light irradiation. Water Res 42:356–362

Kim YJ, Yang SI, Ryu JC (2010) Cytotoxicity and genotoxicity of nano-silver in mammalian cell lines. Mol Cell Toxicol 6:119–125

Kora AJ, Arunachalam J (2011) Assessment of antibacterial activity of silver nanoparticles on *Pseudomonas aeruginosa* and its mechanism of action. World J Microb Biot 27:1209–1216

Korsvik C, Patil S, Seal S, Self WT (2007) Superoxide dismutase mimetic properties exhibited by vacancy engineered ceria nanoparticles. Chem Commun 10:1056–1058

Laban G, Nies LF, Turco RF, Bickham JW, Sepúlveda MS (2010) *The effects of silver nanoparticles on fathead minnow (*Pimephales promelas*) embryos*. Ecotoxicology 19:185–195

Lee HY, Park HK, Lee YM, Kim K, Park SB (2007) A practical procedure for producing silver nanocoated fabric and its antibacterial evaluation for biomedical applications. Chem Commun 28:2959–2961

Lesser MP (2006) Oxidative stress in marine environments: biochemistry and physiological ecology. Annu Rev Physiol 68:253–278

Li T, Albee B, Alemayehu M, Diaz R, Ingham L, Kamal S, Rodriguez M, Bishnoi SW (2010a) *Comparative toxicity study of Ag, Au, and Ag-Au bimetallic nanoparticles on* Daphnia magna. Anal Bioanal Chem 398:689–700

Li WR, Xie XB, Shi QS, Zeng HY, OU-Yang YS, Chen YB (2010b) *Antibacterial activity and mechanism of silver nanoparticles on* Escherichia coli. Appl Microbiol Biotechnol 85:1115–1122

Liau SY, Read DC, Pugh WJ, Furr JR, Russell AD (1997) Interaction of silver nitrate with readily identifiable groups: relationship to the antibacterial action of silver ions. Lett Appl Microbiol 25:279–283

Liu Y, He L, Mustapha A, Li H, Hu ZQ, Lin M (2009) *Antibacterial activities of zinc oxide nanoparticles against* Escherichia coli *O157:H7*. J Appl Microbiol 107:1193–1201

Lok CN, Ho CM, Chen R, He QY, Yu WY, Sun HZ, Tam PKH, Chiu JF, Che CM (2006) Proteomic analysis of the mode of antibacterial action of silver nanoparticles. J Proteome Res 5:916–924

Lok CN, Ho CM, Chen R, He QY, Yu WY, Sun H, Tam PKH, Chiu JF, Che CM (2007) Silver nanoparticles: partial oxidation and antibacterial activities. J Biol Inorg Chem 12:527–534

Luk KFS, Maki AH, Hoover RJ (1975) Studies of heavy-metal binding with polynucleotides using optical detection of magnetic-resonance. Silver(I)binding. J Am Chem Soc 97:1241–1242

Marambio-Jones C, Hoek EMV (2010) A review of the antibacterial effects of silver nanomaterials and potential implications for human health and the environment. J Nanopart Res 12:1531–1551

Martínez-Castañón GA, Niño-Martínez N, Martínez-Gutierrez F, Martínez-Mendoza JR, Ruiz F (2008) Synthesis and antibacterial activity of silver nanoparticles with different sizes. J Nanopart Res 10:1343–1348

Morgan IJ, Henry RP, Wood CM (1997) The mechanism of acute silver nitrate toxicity in freshwater rainbow trout (*Oncorhynchus mykiss*) is inhibition of gill Na^+ and Cl^- transport. Aquat Toxicol 38:145–163

Morones JR, Elechiguerra JL, Camacho A, Holt K, Kouri JB, Ramírez JT, Yacaman MJ (2005) The bactericidal effect of silver nanoparticles. Nanotechnology 16:2346–2353

Moutin MJ, Abramson JJ, Salama G, Dupont Y (1989) Rapid Ag^+-induced release of Ca^{2+} from sarcoplasmic reticulum vesicles of skeletal muscle: a rapid filtration study. Biochim Biophys Acta 984:289–292

Mueller NC, Nowack B (2008) Exposure modeling of engineered nanoparticles in the environment. Environ Sci Technol 42:4447–4453

Nebeker AV, McAuliffe CK, Mshar R, Stevens DG (1983) Toxicity of silver to steelhead and rainbow trout, fathead minnows and *Daphnia magna*. Environ Toxicol Chem 2:95–104

Orrenius S, McCabe MJ, Nicotera P (1992) Ca^{2+}-dependent mechanisms of cytotoxicity and programmed cell death. Toxicol Lett 64–65:357–364

Pal S, Tak YK, Song JM (2007) Does the antibacterial activity of silver nanoparticles depend on the shape of the nanoparticle? A study of the Gram-negative bacterium *Escherichia coli*. Appl Environ Microb 73:1712–1720

Ratte HT (1999) Bioaccumulation and toxicity of silver compounds: a review. Environ Toxicol Chem 18:89–108

Ringwood AH, McCarthy M, Bates TC, Carroll DL (2010) The effects of silver nanoparticles on oyster embryos. Mar Environ Res 69:S49–S51

Robertson JD, Orrenius S (2000) Molecular mechanisms of apoptosis induced by cytotoxic chemicals. Crit Rev Toxicol 30:609–627

Roh JY, Sim SJ, Yi J, Park K, Chung KH, Ryu DY, Choi J (2009) Ecotoxicity of silver nanoparticles on the soil nematode *Caenorhabditis elegans* using functional ecotoxicogenomics. Environ Sci Technol 43:3933–3940

Sakon S, Xue X, Takekawa M, Sasazuki T, Okazaki T, Kojima Y, Piao JH, Yagita H, Okumura K, Doi T, Nakano H (2003) NF-κB inhibits TNF-induced accumulation of ROS that mediate prolonged MAPK activation and necrotic cell death. EMBO J 22:3898–3909

Sanpui P, Chattopadhyay A, Ghosh SS (2011) Induction of apoptosis in cancer cells at low silver nanoparticle concentrations using chitosan nanocarrier. ACS Appl Mater Interfaces 3:218–228

Schreurs WJA, Rosenberg H (1982) Effect of silver ions on transport and retention of phosphate by *Escherichia coli*. J Bacteriol 152:7–13

Shrivastava S, Bera T, Roy A, Singh G, Ramachandrarao P, Dash D (2007) Characterization of enhanced antibacterial effects of novel silver nanoparticles. Nanotechnology 18:225103

Sies H (1997) Oxidative stress: oxidants and antioxidants. Exp Physiol 82:291–295

Silver S, Phung LT, Silver G (2006) Silver as biocides in burn and wound dressings and bacterial resistance to silver compounds. J Ind Microbiol Biot 33:627–634

Simon HU, Haj-Yehia A, Levi-Schaffer F (2000) Role of reactive oxygen species (ROS) in apoptosis induction. Apoptosis 5:415–418

Sondi I, Salopek-Sondi B (2004) Silver nanoparticles as antimicrobial agent: a case study on *E. coli* as a model for Gram-negative bacteria. J Colloid Interf Sci 275:177–182

Stohs SJ, Bagchi D (1995) Oxidative mechanisms in the toxicity of metal ions. Free Radical Bio Med 18:321–336

Tarnuzzer RW, Colon J, Patil S, Seal S (2005) Vacancy engineered ceria nanostructures for protection from radiation-induced cellular damage. Nano Lett 5:2573–2577

The Royal Society and The Royal Academy of Engineering (2004) Nanoscience and nanotechnologies: opportunities and uncertainties. The Royal Society and The Royal Academy of Engineering, London

Thiel J, Pakstis L, Buzby S, Raffi M, Ni C, Pochan DJ, Shah SI (2007) Antibacterial properties of silver-doped titania. Small 3:799–803

Tolaymat TM, El Badawy AM, Genaidy A, Scheckel KG, Luxton TP, Suidan M (2010) An evidence-based environmental perspective of manufactured silver nanoparticle in syntheses and applications: a systematic review and critical appraisal of peer-reviewed scientific papers. Sci Total Environ 408:999–1006

Viarengo A, Nicotera P (1991) Possible role of Ca^{2+} in heavy metal cytotoxicity. Comp Biochem Phys C 100:81–84

Wei LN, Tang JL, Zhang ZX, Chen YM, Zhou G, Xi TF (2010) Investigation of the cytotoxicity mechanism of silver nanoparticles *in vitro*. Biomed Mater 5:044103

Wijnhoven SWP, Peijnenburg WJGM, Herberts CA, Hagens WI, Oomen AG, Heugens EHW, Roszk B, Bisschops J, Gosens I, van de Meent D, Dekkers S, de Jong WH, van Zijverden M, Sips AJAM, Geertsma RE (2009) Nano-silver—a review of available data and knowledge gaps in human and environmental risk assessment. Nanotoxicology 3:109–138

Woodrow Wilson International Center for Scholars (2011) The project on emerging nanotechnologies. http://www.nanotechproject.org/inventories/consumer/analysis_draft/. Accessed 20 Jul 2011

Xia T, Kovochich M, Brant J, Hotze M, Sempf J, Oberley T, Sioutas C, Yeh JI, Wiesner MR, Nel AE (2006) Comparison of the abilities of ambient and manufactured nanoparticles to induce cellular toxicity according to an oxidative stress paradigm. Nano Lett 6:1794–1807

Xia T, Kovochich M, Liong M, Mädler L, Gilbert B, Shi HB, Yeh JI, Zink JI, Nel AE (2008) Comparison of the mechanism of toxicity of zinc oxide and cerium oxide nanoparticles based on dissolution and oxidative stress properties. ACS Nano 2:2121–2134

Yamane T, Davidson N (1962) On the complexing of deoxyribonucleic acid by silver(I). Biochim Biophys Acta 55:609–621

Yeo MK, Kang M (2008) Effects of nanometer sized silver materials on biological toxicity during zebrafish embryogenesis. B Kor Chem Soc 29:1179–1184

Yu BP (1994) Cellular defenses against damage from reactive oxygen species. Physiol Rev 74:139–162

Zhao CM, Wang WX (2010) Biokinetic uptake and efflux of silver nanoparticles in *Daphnia magna*. Environ Sci Technol 44:7699–7704

Zhao CM, Wang WX (2011) Comparison of acute and chronic toxicity of silver nanoparticles and silver nitrate to *Daphnia magna*. Environ Toxicol Chem 30:885–892

Zimmermann R, Wickner W (1983) Energetics and intermediates of the assembly of protein OmpA into the outer membrane of *Escherichia coli*. J Biol Chem 258:3920–3925

Diazinon—Chemistry and Environmental Fate: A California Perspective

Vaneet Aggarwal, Xin Deng, Atac Tuli, and Kean S. Goh

Contents

1 Introduction 107
2 Physicochemical Properties 108
3 Use Profile for Diazinon in California 110
4 Environmental Fate 114
    4.1 Soil and Sediment 114
    4.2 Water 122
    4.3 Air and Precipitation 126
5 Aquatic Toxicology 129
    5.1 Mode of Action 129
    5.2 Bioaccumulation 132
    5.3 Aquatic Life Benchmarks and Water Quality Criteria 133
6 Summary 134
References 135

1 Introduction

Diazinon (O,O-diethyl O-2-isopropyl-6-methylpyrimidin-4-yl phosphorothioate) was first registered in the USA in 1956 (US EPA 2006) by the Swiss company J.R. Geigy. Diazinon is a broad-spectrum contact organophosphorus pesticide that is used as an insecticide, acaricide, and nematicide. Diazinon has been widely used to control soil and foliage insects and pests on a wide range of crops such as rice, fruits, wine grapes, sugarcane, corn, and potatoes. Diazinon is also used to control mange mites, ticks, lice, biting flies on sheep, cows, pigs, goats, and horses. In California, diazinon has been applied primarily on fruits, vegetables, and for

V. Aggarwal (✉) • X. Deng • A. Tuli • K.S. Goh
Department of Pesticide Regulation, California Environmental Protection Agency,
1001 I Street, Sacramento, CA 95812-4015, USA
e-mail: vaggarwal@cdpr.ca.gov

Fig. 1 Chemical structure of diazinon

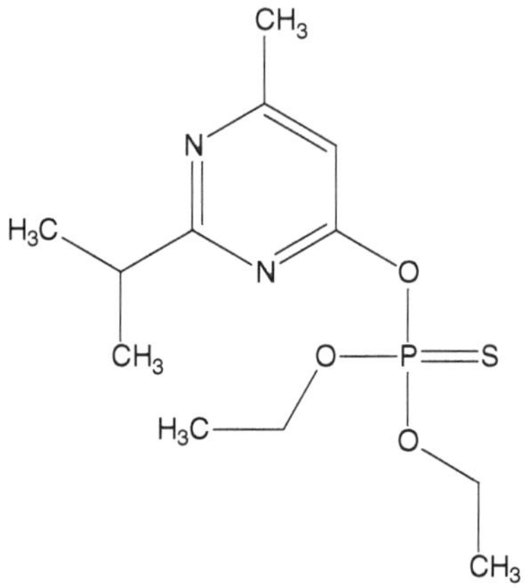

landscape maintenance and structural pest control. In 2010, a total of only 64,122 kg of diazinon was used in California (CDPR 2010a). Diazinon was formerly used in household and garden products for pest control. However, manufacturing of indoor use products was discontinued on June 30, 2001, and production of nonagricultural outdoor use products containing diazinon was discontinued on June 30, 2003. As of December 31, 2004, sales of diazinon-containing products for residential use ceased (US EPA 2000), resulting in diazinon falling to a rank of 94 among the most used pesticides in California by 2010 (CDPR 2010b). Diazinon is formulated as a wettable powder, granules, liquid concentrates, seed dressings, microencapsulations, and impregnated materials (US EPA 2006). Some typical formulations containing diazinon as an active ingredient (a.i.) include the following: Basudin® 10 (10% a.i.), Knoxout® (Pennwalt, 23% a.i.), Nucidol® 60 (60% a.i.), Alfatox®, Gardentox®, and several other trade-named products. Although concentrations have decreased nationwide in urban waters, diazinon is still frequently detected as a residue in agricultural watersheds. In California's agricultural regions with the highest diazinon use, the detection frequencies in 2005–2010 could reach 90% of the monitoring samples; moreover, the exceedance rate vs. the water quality criterion of 0.1 µg/L was 66.7% (Zhang and Starner 2011). This large proportion of detections and exceedance of water quality criteria have led to concerns about diazinon's potential environmental impacts. In this chapter, we provide a review of the environmental fate of diazinon and describe its toxicity to aquatic organisms.

2 Physicochemical Properties

Diazinon (Fig. 1), an organophosphorus pesticide, is a colorless to dark brown liquid, and is denser than water (1.116 g/cm^3). Chemical Abstracts Service (CAS) registry number of diazinon is 333-41-5. In addition to the International Union of

Table 1 Physicochemical properties of diazinon

CAS Registry Number	333-41-5
Chemical name (CAS)	*O,O*-diethyl *O*-[6-methyl-2-(1-methylethyl)-4-pyrimidinyl] phosphorothioate
Chemical name (IUPAC)	*O,O*-diethyl *O*-2-isopropyl-6-methylpyrimidin-4-yl phosphorothioate
Chemical formula	$C_{12}H_{21}N_2O_3PS$
Molecular weight	304.35 g/mol
Physical form	Colorless to dark brown liquid
Water solubility	0.06 g/L (20°C)[a]
	0.054 and 0.069 g/L (20–40°C)[b]
Density	1.116 g/cm^3 (20°C)[c, d]
	1.118 (4°C)[c]
Octanol/water partition coefficient, Log K_{ow}	3.69 (pH 7, 20°C)[a]
	3.29[c]
	3.3–3.81[b]
Aqueous photolysis half-life	5.05 days (pH 7, 25°C)[e]
	50 days (pH 7)[a]
Hydrolysis half-life	12.4 days (pH 5, 24°C)[e]
	43.3 days (pH 7.4, 16°C)[f]
	138 days (pH 7, 24°C)[e]
Vapor pressure	8.4×10^{-5} mmHg (20°C)[b]
	8.97×10^{-5} mmHg (25°C)[a, e]
	2.8×10^{-4} mmHg (30°C)[g]
Boiling point	82–84°C (2.0×10^{-4} mmHg)[h]
K_{oc}	40–432 mL/g (mean 191) OC depending on soil type and environmental conditions[b]
Henry's law constant	1.4×10^{-6} atm.·m^3/mol (25°C)
	1.13×10^{-7} atm.·m^3/mol depending on the technique used[b]
	6.01×10^{-7} atm.·m^3/mol (25°C)[a]

[a] IUPAC (2010)
[b] US ATSDR (2008)
[c] Budavari et al. (1989)
[d] Mackay et al. (2006)
[e] CDPR (2010a)
[f] Morgan (1976)
[g] Lichtenstein and Schulz (1970)
[h] Gysin and Margot (1958)

Pure and Applied Chemistry (IUPAC) name given in the introduction, the CAS name for diazinon is *O,O*-diethyl *O*-[6-methyl-2-(1-methylethyl)-4-pyrimidinyl] phosphorothioate. Diazinon has a boiling point of 82–84°C and relatively high water solubility of 60 mg/L. Pesticides with high water solubility may leach below the root zone and at times reach groundwater or can be dispersed by surface runoff far from their application site (Leonard 1990; Wolfe et al. 1990). The major physicochemical properties for diazinon are summarized in Table 1.

3 Use Profile for Diazinon in California

California is the top agricultural state in the USA and grows more than half of the nation's fruits, vegetables, and nuts. Moreover, California is the only state with an extensive pesticide use reporting system, which was initiated in 1990. Under the pesticides use reporting program, all agricultural pesticide use in all 58 California counties (Fig. 2) must be reported monthly to corresponding county agricultural

Fig. 2 California counties

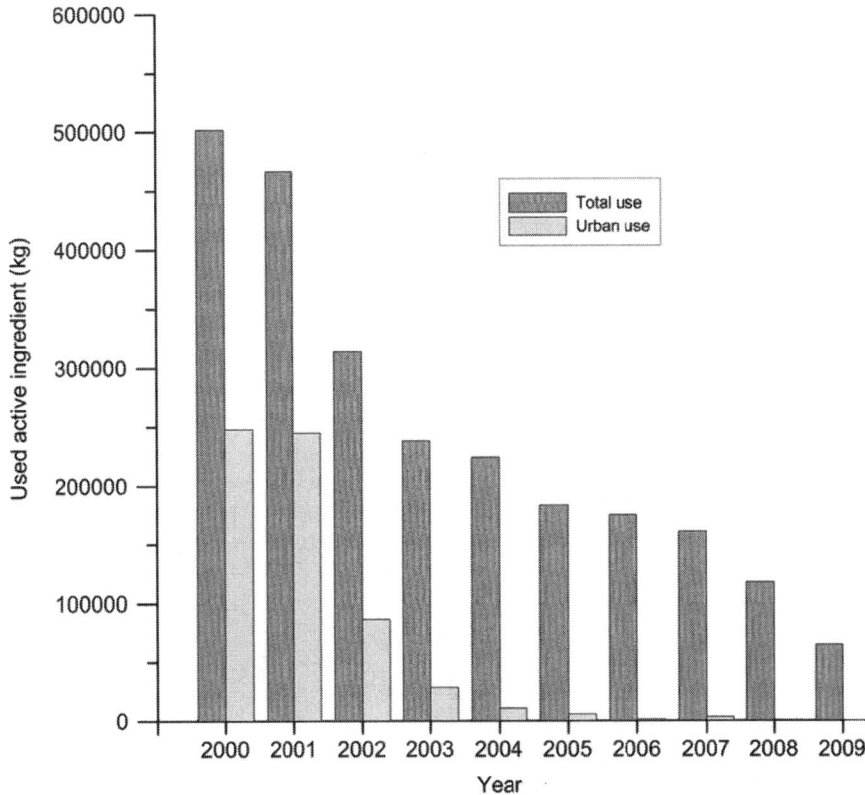

Fig. 3 Reported total and urban use of diazinon active ingredient from 2000 to 2009 in California

commissioners, who, in turn, report their data to Department of Pesticide Regulation (DPR). Consumer home and garden uses are not included in this use profile, but professional landscape maintenance, commercial structures, parks, etc. applications are reported. From 2000 to 2009, a total of 2,445,277 kg of diazinon was reported to have been used in California (CALPIP 2011). There was a significant reduction in urban use (Fig. 3) of diazinon that resulted from a December 2000 agreement between the technical product registrant and the US EPA; this agreement was driven by EPA action to phase out and cancel all indoor and outdoor residential uses of diazinon, to reduce risks to children and others (US EPA 2004).

Diazinon use among the top ten counties in California from 2000 to 2009 ranged from 69,648 kg a.i. (San Benito County) to 619,074 kg a.i. (Monterey County) (Table 2). Los Angeles and Stanislaus Counties had the largest decrease in diazinon use, because a majority of their use was for applications to urban areas. The other counties in the diazinon top ten list predominantly used diazinon in agricultural settings (Table 3). Treated acreage remained steady for the period 2000–2007. Although the total amount of diazinon applied (Fig. 3) decreased considerably from the phase out of outdoor and indoor residential uses, the agricultural use of diazinon remained nearly constant until 2008. The effect of phasing diazinon out of agricultural use

Table 2 Total diazinon use (kg a.i.) by the top ten counties in California during the period 2000–2009

County	Year										Total
	2000	2001	2002	2003	2004	2005	2006	2007	2008	2009	
Monterey	56,316	64,773	65,061	72,554	78,202	73,552	65,813	65,810	53,673	23,319	619,074
Los Angeles	102,926	113,420	11,891	6,691	2,721	674	2,048	1,010	2,493	683	244,557
Fresno	38,731	32,622	26,598	19,384	19,426	18,894	23,255	19,846	12,674	4,716	216,146
Imperial	21,840	18,472	20,383	19,937	18,796	14,504	13,949	11,592	5,650	2,390	147,513
Stanislaus	33,400	27,998	11,602	6,516	7,293	3,610	3,836	2,710	1,106	1,655	99,727
Kern	17,845	14,396	16,898	7,222	5,122	6,969	6,089	9,383	4,859	2,968	91,751
Sutter	12,897	9,732	13,581	11,747	7,287	6,415	9,805	6,579	4,594	5,371	88,007
Tulare	23,327	19,801	13,728	6,167	4,965	3,511	2,920	1,397	1,008	1,064	77,890
Santa Clara	11,033	13,603	14,020	6,292	7,610	6,269	5,974	6,709	3,363	2,016	76,890
San Benito	10,949	9,058	8,144	6,805	7,633	6,478	6,935	6,478	3,960	3,208	69,648

Table 3 Distribution of urban and agricultural uses (kg a.i.) of diazinon in the top ten California counties between 2000 and 2009

County	Urban	Agricultural	Total
Monterey	4,735	614,339	619,074
Los Angeles	216,426	28,131	244,557
Fresno	20,978	195,169	216,146
Imperial	883	146,630	147,513
Stanislaus	58,258	41,469	99,727
Kern	5,713	86,038	91,751
Sutter	967	87,040	88,007
Tulare	34,144	43,746	77,890
Santa Clara	30,130	46,761	76,890
San Benito	920	68,727	69,648

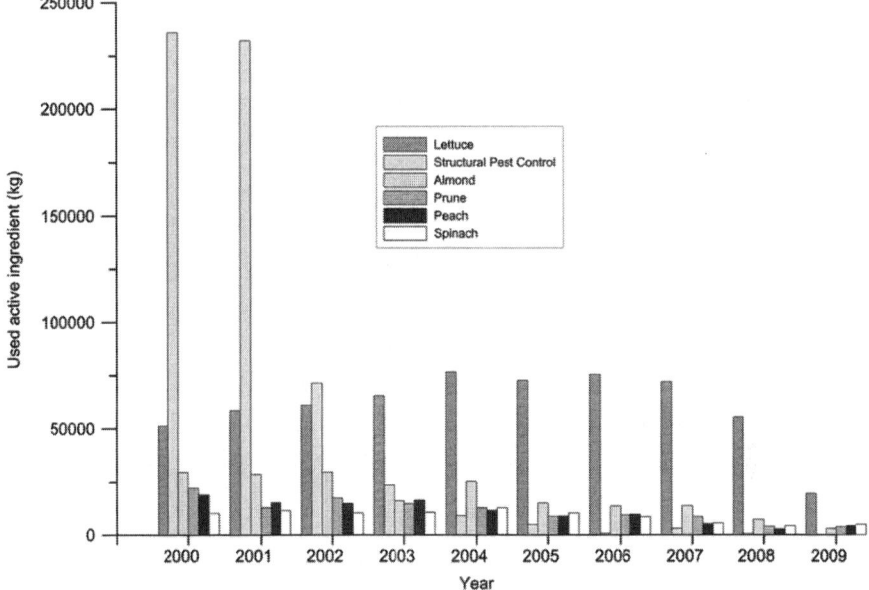

Fig. 4 Total diazinon use (kg a.i.) by the top six crop uses in California during the period 2000–2009

was more pronounced in 2009; in 2009, the total treated area was reduced to 56,905 ha, only 55 and 67% of that treated in 2008 and 2007, respectively.

In Fig. 4, we present the total amount of diazinon-active ingredient applied to the top six crops in California. Overall, lettuce had the highest diazinon use (608,265 kg) in the years 2000–2009 (Fig. 4). Although structural pest control had the highest use of diazinon in 2000–2002, this use sharply decreased after 2003, and in 2009, diazinon use for structural pest control was largely phased out (83 kg) (Fig. 4). The other top ranking crops were almond, prune, peach, and spinach. Diazinon was applied primarily by ground equipment from 2000 to 2009, followed by aerial application (Fig. 5).

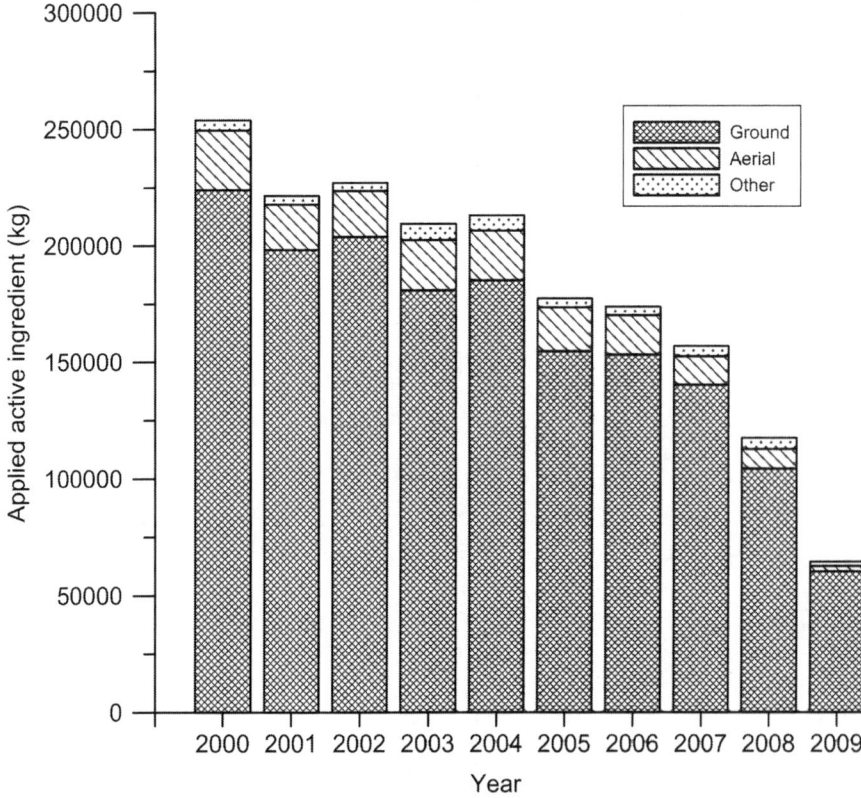

Fig. 5 Total diazinon use (kg a.i.) by application types in California during the period 2000–2009

4 Environmental Fate

4.1 Soil and Sediment

Contamination of surface and groundwater by pesticides is a major concern. Once released to soils, diazinon can be adsorbed by the soil, degraded by several processes, including hydrolysis, photolysis, and microbial degradation, or may move away from the application site by leaching, volatilization, etc. Microbial degradation is generally the major pathway by which diazinon is degraded in soils. Diazinon is moderately mobile as a result of its low to moderate sorption capacity to soils (K_{oc} values range from 40 to 432, mean value 191). Sorption by soil plays an important role in the protection of surface and groundwater from contamination by diazinon, and such studies have been conducted to study the sorption–desorption behavior of diazinon by soils. Below, we present a summary of some of these sorption studies.

4.1.1 Sorption

Diazinon adsorption to soils depends on the soil pH, organic matter (OM) content, and clay content. Adsorption generally increases with increases in OM and clay content (Armstrong et al. 1967). The tendency of a pesticide to bind to soil or sediment particles is often characterized by its organic carbon-normalized soil adsorption coefficient (K_{oc}). Pesticides with relatively high K_{oc} values tend to remain in the soil or attached to soil particles entrained in flowing water, restricting or slowing their movement downstream. Pesticides that have relatively low K_{oc} values tend to bind less tightly to soil particles, and therefore are likely to be leached from the soil and transported by moving water. K_{oc} values have been determined for diazinon in a variety of soil types and have been reported to range between 40 and 432 (mean of 191). The range of these values indicates that diazinon has a low to moderate tendency to remain bound to soil and sediment, and therefore is moderately mobile.

Konrad et al. (1967) equilibrated diazinon with three soils (loamy sand, silty clay loam, and clay) and observed an initial rapid decrease in diazinon concentration from soil adsorption, that was followed by a slow but continual decrease in concentration, which they attributed to degradation. In 24 h, diazinon degradation was 11, 7, and 6% in the silty clay loam, clay, and loamy sand soils, respectively. After 10 h, the release of degradation products approached linearity (measured by release of degradation products), and this release was related to rates of initial soil adsorption. Silty clay loam soil had the highest OM content and the highest diazinon sorption of the three soils studied. Armstrong et al. (1967) reported that adsorption should increase with increasing OM and clay content, but was more closely related to OM than to clay content. Thus, the greater adsorption in the silty clay loam (10.0% OM) over clay (3.8% OM) and loamy sand (1.6% OM) soils was likely caused by the higher OM content of the silty clay loam soil. Konrad et al. (1967) further stated that if the mechanism of diazinon adsorption is mainly through H-bonding, then a decrease in soil pH should increase the extent of adsorption. They observed similar adsorption in both clay and loamy sand soils and concluded that the greater acidity of loamy sand (pH 3.8 vs. 6.4 in clay soil) soil compensated for the slightly greater OM content (3.8% OM in clay vs. 1.6% OM in loamy sand soil) and much greater clay content of the clay soil (18.7% in clay soil and 5.2% in loamy sand soil). They further reported that the effect of higher OM content of the silty clay loam soil could not be compensated for by either the higher clay content of the clay soil or the lower pH of the loamy sand soil.

Arienzo et al. (1994b) observed that more diazinon was sorbed by two soils (loamy sand and sandy loam), when these soils were modified with three organic amendments and with one carbon-rich organic compound, hexadecyltrimethylammonium bromide (HDTMA). HDTMA-modified soil retained the most diazinon amongst all four treatments, retaining as much as 70.5% of the applied diazinon (5 mL of 200 µg/mL, equivalent to 5 kg/ha), whereas control soil only retained 25.2% diazinon. The amounts retained by the soils treated with the other three amendments were ~30% in sandy loam soil and about 25% in loamy sand soil. These small differences in the amount of retained diazinon can be attributed to small differences in the clay and OM content of the two soils. The higher sorption by HDTMA-treated soil may be related to the OM content of the treated soils, as HDTMA had the highest organic carbon content of the four amendments. In another study, Arienzo et al. (1994a) studied diazinon adsorption by 25 soils and found soil OM content to be the most important factor in diazinon's sorption in soils containing over 2% OM. They also reported that diazinon adsorbs to soils that have OM content below 2%, and the adsorption was influenced by the silt and clay soil components. Other authors have reported that no correlation existed between soil OM content at content levels below 2% (Calvet et al. 1980; Sanchez-Camazano and Sanchez-Martin 1984; Sanchez-Martin and Sanchez-Camazano 1991). Arienzo et al. (1993) conducted a study on adsorption and mobility of diazinon in soils with aqueous media and mixtures of methanol–water and hexane–water. They reported decreased diazinon adsorption in hexane–water and methanol–water systems, whereas increasing soil OM content increased diazinon adsorption in aqueous media. This shows that diazinon has higher affinity for an organic solvent than soil.

Sarmah et al. (2009) studied diazinon sorption in two New Zealand soils at two depths (0–10 cm and 40–50 cm), and observed that diazinon adsorption in the surface layer (27.7 L/kg) was much higher than in the deeper layer (3.7 L/kg); they attributed this difference to higher organic carbon content of the surface layers. It has been reported that organic matter can bind a substantial proportion of pesticide residues present during the composting process (Frederick et al. 1996; Petruska et al. 1985). For example, after 21 days of composting, a substantial amount of ^{14}C-radiolabeled diazinon was associated with humic and fulvic acids (15%), and with the humin fraction (25%) (Petruska et al. 1985). These results indicate that diazinon mobility in soils is reduced by organic matter, and the nature and composition of organic compounds added to soil, along with the carbon content, have a profound effect on diazinon movement in soils.

Nemeth-Konda et al. (2002) estimated octanol–water partition coefficients (P_{ow}) by using either a computer model or high performance liquid chromatographic parameters, and observed good agreement between experimental and the computer-model estimated data. The octanol–water partition coefficient (P_{ow}) is a useful parameter for predicting the soil adsorption behavior of a chemical. Diazinon had the highest log P_{ow} (3.86) value among six pesticides studied, namely, acetochlor, atrazine, carbendazim, diazinon, imidacloprid, and isoproturon, indicating that diazinon was the most hydrophobic of the group (Nemeth-Konda et al. 2002). These researchers also conducted sorption studies using Hungarian sandy loam soil and observed that the percentage adsorbed by the soil decreased as the initial concentration of diazinon increased from 0.04 to 5.0 mg/L. These authors further conducted desorption experiments immediately after adsorption to study the intensity of diazinon–soil interaction, and observed that only about 60% of the sorbed amount desorbed, concluding that there is a strong bonding between diazinon and the soil. They also concluded that the higher adsorption capacity of diazinon may be related to diazinon's low water solubility and high hydrophobicity, and thus the rate of diazinon's migration to groundwater will be slower than that of the other six pesticides studied.

4.1.2 Leaching

As previously mentioned, diazinon has a K_{oc} value of 40–432 (mean of 191) (US ATSDR 2008), and is considered to be moderately mobile in soil (US EPA 2006). Among the factors that influence diazinon's leaching potential is soil type (e.g., clay vs. sand), amount of rainfall, depth of the groundwater, and extent of degradation. In laboratory tests of sandy and organic soils, Sharom et al. (1980a) used 10 and 4 g of diazinon-treated (2 ppm) sand and muck, respectively, and found that the amount of diazinon leached from sand decreased with each successive 200-mL wash. They reported that 95% of the added diazinon leached from the sand after ten successive 200-mL washes. However, only 50% leached from the organic soil. Arienzo et al. (1994b) examined the mobility of diazinon in two soils (loamy sand and sandy loam). Each soil was treated with one of four organic amendments (two types of peat, liquid humic amendment, and hexadecyltrimethylammonium bromide (HDTMA)). The authors observed that all four amendments reduced the leaching of diazinon from both soils, with HDTMA being the most effective. The cumulative amount of diazinon in the leachate decreased from 49.5% in the control soil to 18.3% in the HDTMA-treated soil. These results indicate that diazinon mobility in soils is reduced by adding organic materials to soil. The nature and composition of organic compounds added to soil, along with the carbon content, have a profound effect on diazinon movement in soils.

Arienzo et al. (1994a) tested the adsorption and mobility of diazinon in 25 soils having different physicochemical properties. Diazinon movement was correlated to soil organic matter content; diazinon was found to be slightly mobile in soils with organic matter content < 3%, and immobile in soils with organic matter content > 3%. Levanon et al. (1994) studied the impact of plow tillage on microbial activity in the top 5-cm soil layer, and observed a higher leaching rate for diazinon in plow tillage soils (20.7%) than in no-tillage soils (2.8%), after incubation for 21 days. They attributed the lower diazinon leaching in no-tillage soils to a higher diazinon mineralization rate in these soils. The no-tillage soils were characterized by having a higher organic matter content and higher microbial populations and activity than did plow tillage soils. In another experiment, Arienzo et al. (1993) conducted a soil adsorption and mobility study on diazinon with aqueous media and mixtures of methanol–water and hexane–water. They reported that increasing the soil organic matter content increased diazinon adsorption in aqueous media, whereas an increase in organic solvent concentration decreased diazinon adsorption. For example, the value of the Freundlich adsorption parameter (K) decreased from 8.79 in water to 6.64 and 3.10 in 10 and 20% methanol–water systems, respectively. This decrease was more drastic in the hexane–water system, with K values decreasing to 1.53 and 0.31 in 10 and 20% hexanol–water systems, respectively. The co-presence of such organic solvents with pesticide waste residues, such as at hazardous waste disposal sites, was postulated by the authors to increase the potential for groundwater contamination by increasing the downward movement of diazinon.

Ritter et al. (1973) studied the effect of soil temperature, soil moisture, and soil bulk density on the diffusion of diazinon, atrazine, and propachlor in a silt loam soil under laboratory conditions. They reported that the greatest movement of diazinon occurred at high soil temperatures, while soil moisture content and soil bulk density had very little effect on diazinon movement.

Leistra et al. (1984), using a computer model of a simplified soil system, simulated the downward flow of water and movement of diazinon through the unsaturated zone in glasshouse soils. They observed that under average situations (irrigation rate, adsorption coefficients, and transformation rate), downward movement of diazinon was restricted and concentrations below 15 cm remained very low. When soils with comparatively low organic matter were used, or the rate of irrigation was increased, diazinon moved deeper in the soil horizon, although most of it was retained in the upper 20 cm soil. Ritter et al. (1973) reported that the degradation of diazinon increased with increased soil temperature and moisture content. They reported that at 43°C, more than 96% of diazinon degraded within 8 days, while 75% of diazinon degraded at 10°C in the same period. They also observed that at 8% soil moisture content, 15.6% diazinon remained in the soils after 8 days, while at 23% soil moisture content, only 12.5% diazinon remained in the soil in the same time period (Ritter et al. 1973). In another study, Ritter et al. (1974) observed only small concentrations (<1.0 ppm) of diazinon in soil samples a few hours after diazinon application, and did not detect any diazinon 21 days after diazinon application. These results show that diazinon degrades rapidly in soil. Consequently, leaching is a relatively unimportant fate process for diazinon.

4.1.3 Mineralization

Mineralization is defined as the process by which organic substances are converted to inorganic substances. Mineralization of a pesticide eventually reduces it to carbon dioxide, water, and other inorganic components. Higher mineralization rates indicate that microorganisms are using an organic substance as a source of carbon and energy. Fenlon et al. (2007) studied diazinon's mineralization in five soils (four organically managed and one conventionally managed) that were aged for 14 weeks. They observed appreciable mineralization of diazinon in two of the organically managed soils. These soils had the highest OM content of the five studied soils. Petruska et al. (1985) used radiolabeled diazinon to study the fate of diazinon under composting systems. Their results indicated that a substantial fraction of diazinon was entrapped within organic matter in a form that resisted extraction by organic solvents. Further, they reported that only 0.2% diazinon mineralized after 3 weeks of composting. Similar results were obtained by Leland et al. (2003), who reported that only 0.2% of the applied ^{14}C-diazinon was mineralized to CO_2 over 30- and 60-day incubation time periods. The low diazinon mineralization rates observed in these studies indicate that diazinon is not used as a carbon and energy source by soil microorganisms.

Levanon et al. (1994) studied the effects of plow tillage on microbial activity and diazinon degradation in the top (0–5 cm) soil layer, and observed higher mineralization rates of diazinon in no-tillage soils (40% mineralization vs. 20% in plow-tillage soils, after 30 days). They attributed this increase in mineralization rates to higher microbial populations and activity in no-tillage soils, which resulted in decreased leaching from soils with no-tillage vs. plow-tilled soils. Microbial population and activity, measured as biomass, bacterial counts, hyphal length of fungi, and carbon dioxide evolution, were all lower in samples of plow tillage soils.

4.1.4 Dissipation

Dissipation of pesticides relates directly to their persistence (or length of time they reside in the environment). Generally, pesticides remaining in the field for many weeks after application are of concern, and are more prone to move because of their environmental longevity. The field dissipation half-life of a pesticide is a value that expresses the time required for half of a given quantity to degrade or dissipate from the soil. Diazinon has a low persistence in soil, with a half-life of only 34.8 days (Singh and Singh 2005). In a field study conducted to monitor the changes in diazinon concentration over time, Ando et al. (1993) detected the highest diazinon soil residues (17.1 µg/g) immediately after the first diazinon application; these amounts decreased to 0.38 µg/g 3 days after the final application. Based on these results, they concluded that diazinon dissipates quickly from the soil. Sarmah et al. (2009) reported that both depth and temperature had an effect on the dissipation half-life (DT_{50}) of diazinon, with lower layers having significantly higher DT_{50} values. They also observed that a decrease in temperature from 20 to 7.5°C caused a threefold increase in the subsoil half-life (25 days—Waikiwi soil; 12 days—Motupiko soil) as compared to surface soil (7 days—Waikiwi soil; 5 days—Motupiko soil). Sattar (1990) incubated silty clay (OC = 0.6%) and sandy clay (OC = 0.5%) soils at 25°C, and reported half-life values for diazinon of 28 days and 36 days, respectively. These soils had lower organic carbon content than the soils used by Sarmah et al. (2009), which may explain the longer half-life values reported by Sattar (1990).

Sarmah et al. (2009) reported that organic carbon content and microbial activity decreases with an increase in soil depth. This decrease reduces soil sorption capacity for organic compounds, which, in turn, slows the degradation rate. Sethunathan and Macrae (1969a) investigated the role of soil microflora on the fate of diazinon in flooded Philippine soils. They reported that diazinon disappeared from both sterilized and nonsterilized submerged soils, and followed first-order kinetics. They obtained half-life values of 8.8 and 33.8 days for nonsterilized and sterilized clay soil, respectively, while the corresponding values for clay loam soil were 17.4 and 43.8 days. Based on these results, the authors concluded that soil microflora are involved in the rapid dissipation of diazinon. Getzin and Rosefield (1966) and Gunner et al. (1966) reported that diazinon persists for more than 160 days in nonflooded soils. Levot et al. (2004), on the other hand, reported that diazinon had a half-life of 64 days under waterlogged conditions, and attributed this to slower biodegradation from oxygen depletion under such soils. In another study, Sethunathan and Yoshida (1969) observed that diazinon degraded faster in nonsterilized than sterilized flooded soil, resulting in greater accumulation of the diazinon hydrolysis product, 2-isopropyl-6-methyl-4-hydroxy pyrimidine. These differences were more pronounced in the first 10 days. The authors recovered approximately 40% of radioactive-applied material in both soils and concluded that diazinon or its degradation products could possibly have become bound to the soil during incubation. These results indicate that diazinon persists longer under nonflooded and sterilized soil conditions.

Sethunathan and Macrae (1969a) reported that more than 90% of the applied diazinon was lost 70 days after application in three submerged tropical soils. Degradation was more rapid in two of the three nonsterilized soils, indicating that microbes were involved in diazinon's degradation. In the third acidic clay soil (pH 4.7), diazinon degradation was more rapid in the sterilized samples, and the authors attributed this to the compound's instability under acid conditions. These results suggest that microbial activity plays an important role in diazinon's degradation (Schoen and Winterlin 1987). Szeto et al. (1990) reported that diazinon dissipated rapidly from irrigation ditch sediment (pH 4.4) and from sediment in an adjacent reservoir (pH 5.0). They did not detect diazinon in sediment 66 days after the second diazinon application, whereas 120 ppb of diazinon was still present in irrigation ditches.

Diazinon degraded more rapidly in acidic organic soils (pH 5.2) than in neutral mineral soils (pH of 6.8 and 8.0) (Chapman and Cole 1982). Schoen and Winterlin (1987) studied the effects of various soil factors and organic amendments on diazinon degradation. Diazinon degradation was affected by pH, soil type, organic amendments, soil moisture, and pesticide concentration, with soil pH being a major factor affecting the degradation of diazinon. At a soil concentration of 100 ppm of diazinon, and 50% water saturation, estimated half-life values at pH 4, 7, and 10 were 66, 209, and 153 days, respectively, in sandy loam soil; 49, 124, and 90 days, respectively, in clay loam; and 14, 45, and 64 days, respectively, in sandy loam amended with peat. The authors speculated that soil pH was lowered by the addition of acidic peat to the soil, and this could have been responsible for increased diazinon degradation at pH 10. These results indicate that diazinon degrades rapidly in neutral soils. They also showed that diazinon degradation was slower at high diazinon concentrations in neutral or basic mineral soil, and faster when diazinon was present at low concentrations in moist soil that was amended with peat or acidified to a pH of 4 (Schoen and Winterlin 1987). Felsot et al. (2003) reported that there is a critical diazinon level above which soil diazinon dissipation slows significantly, and he attributed this to slowed degradation from diazinon toxicity to the microorganisms.

4.1.5 Photolysis

Photolysis of diazinon on soil surfaces was studied by Burkhard and Guth (1979). They reported 2-isopropyl-6-methylpyrimidin-4-ol as the main breakdown product (Fig. 6). The photolysis rate was only slightly less on dry soil surfaces (viz., 44% of diazinon degraded in 24 h) than on moist soil surfaces (viz., 51% of diazinon degraded

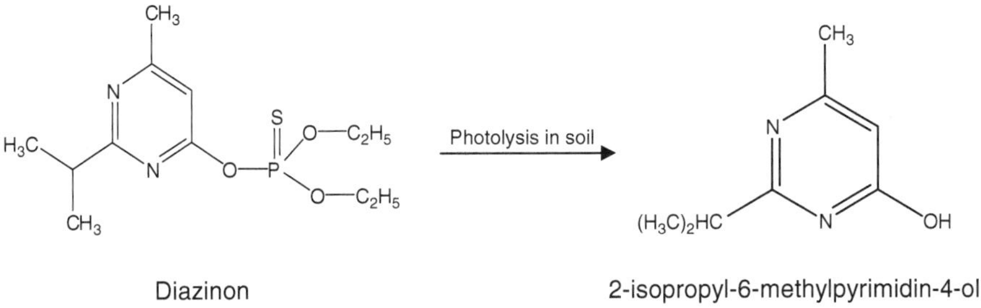

Diazinon → Photolysis in soil → **2-isopropyl-6-methylpyrimidin-4-ol**

Fig. 6 Photolytic breakdown pathway for diazinon on the soil surface (Burkhard and Guth 1979)

in 24 h). Mansour et al. (1997) reported that diazinon degraded more rapidly in river than in distilled water, and the rate of degradation increased with exposure to sunlight. After 14 days, only 5% of applied diazinon remained in river water exposed to sunlight, while more than 80% was present in distilled water kept in the dark.

4.1.6 Effects on Soil Microorganism Populations

Singh and Singh (2005) reported a significant increase in bacteria (14%, $p<0.05$) and azotobacter (27%, $p<0.01$) populations 15 days after diazinon soil treatment. This study was conducted for three consecutive years (diazinon was applied every year), and adverse effects were observed on actinomycetes (10%) and fungal (11%) populations. The authors observed a similar increase in bacteria and azotobacter populations in years 2 and 3 of the study period. The actinomycetes population increased only 60 days after diazinon was applied, by which time initial diazinon residue levels had declined. Sethunathan and Macrae (1969b) also observed a significant increase in the soil actinomycetes population 7 weeks after diazinon soil treatment (two applications, 20 days apart). The actinomycetes population increased from 5,700 (per gram oven dry soil) to 44,800 and 51,100, in soil treated with diazinon at 2 and 20 kg/ha, respectively. Similarly, Gunner et al. (1966) observed an increased actinomycetes population (360,000 vs. 1,000) after diazinon was applied to soil at 3 lb/A. Gunner (1970), in a study with potted bean plants, concluded that the number of microorganisms in the rhizosphere doubled 7 weeks after application of diazinon.

Levanon et al. (1994) studied the effects of plow tillage on microbial activity in the top (0–5 cm) soil layer. Microbial population levels and activity were measured (i.e., biomass, bacterial counts, hyphal length of fungi, and carbon dioxide evolution) and were all higher in no-tillage soil samples. Higher microbial activity resulted in decreased leaching from no-tillage soils, but was associated with a higher rate of diazinon leaching from the plow tillage soils. Further, they observed synergistic effects between fungi and bacteria in the degradation rates of diazinon. The authors noted that almost no mineralization of the compound occurred when either fungi or bacteria were selectively inhibited, demonstrating that synergism existed between the two microbial communities.

Leland et al. (2003) studied the effect of composting on the fate of diazinon and observed that diazinon composted for 30-days was toxic to earthworms; no mortality was observed in earthworms exposed to soil amended with 60-day composted diazinon. The authors reported that 95% of diazinon had degraded after 60-days of composting, and they concluded that composting high concentrations of diazinon greatly reduced the toxicity and amount of diazinon that is bioavailable to soil microorganisms. Ingram et al. (2005) studied the effect of diazinon and imidacloprid on the microbial urease activity in different soils. The studies were conducted with *Proteus vulgaris*, a representative heterotrophic bacteria, and were focused on growth rate. They concluded that increasing the diazinon concentration significantly reduced maximum *P. vulgaris* cell density after 24 h ($p=0.004$). Sethunathan and Macrae (1969b) observed a visible increase in algal populations in the standing water of the diazinon-treated soil. The authors concluded that diazinon generally has a synergistic effect on soil microorganism populations, and quickly dissipates from soil.

4.2 Water

4.2.1 Environmental Monitoring

Diazinon residues in the agricultural environment have been monitored in several studies performed in the USA and Canada. In a national surface water quality monitoring study (1976–1980) performed by Carey and Kutz (1985), diazinon residues were detected in sediment in only 0.5% of samples, with a maximum residue detected of 7.1 ppb. Sapozhnikova et al. (2004) collected sediment samples in 2000 and 2001 from the Salton Sea, an agricultural drainage reservoir in California, and reported diazinon concentrations ranging from 0.5 to 5.4 ppb dry wt in the sediments. Wan et al. (1994) monitored ditch water, soils, and sediments in an agricultural area in the lower Fraser River Valley of British Columbia, Canada. They detected diazinon at five sites in the top 0–5 cm soil layer at an average concentration of 219 ppb, whereas diazinon residues were detected in ditch sediments at three sites at an average concentration of 16 ppb. Further, they reported that diazinon was detected in ditch water at seven locations at an average concentration of 0.07 ppb. Diazinon is of concern as a water contaminant because it may appear in surface or groundwater samples that are slated for drinking purposes.

4.2.2 Groundwater

Diazinon has been detected in irrigated agricultural areas where heavy diazinon applications were made in USA (Cohen 1986) and Canada (Frank et al. 1987, 1990). A groundwater contamination study was performed in 28 of California's 58 counties, in which more than 50 pesticides (from both point and nonpoint sources) were monitored; diazinon was detected in 12 samples (Cohen 1986). Similarly, between 1979 and 1984, diazinon was detected in rural wells of southern Ontario, Canada (Frank et al. 1987) and in farm wells monitored between 1986 and 1987 (Frank et al. 1990). Holden (1986) reported that diazinon was detected in California's groundwater at a maximum concentration of 9 μg/L, and Long (1989) detected diazinon in 5 wells located in the vicinity of agrichemical dealer facilities (total 56 wells studied) at a maximum concentration of 1.1 μg/L (the mean concentration was 0.55 μg/L). Similarly, diazinon residues appeared in water from drinking wells in a single south-central Connecticut town that relies on groundwater for its potable water source (Eitzer and Chevalier 1999). Leistra et al. (1984) collected 22 water samples from water courses in areas that had many glasshouses in the Netherlands. They found measurable diazinon concentrations in nine samples, with the two highest values being 8.7 and 21 μg/L. A search of California Department of Pesticide Regulation's groundwater database (CDPR 2011) for the period 1985–2011 indicated that only one well tested positive for diazinon residues (2.6 μg/L). In a follow-up sampling after 2 months, diazinon was observed at a concentration of 0.2 μg/L.

US EPA (2006), using data from the United States Geological Survey National Water Quality Assessment Program (NAWQA), reported diazinon groundwater detections from a variety of sources that included drinking water wells, monitoring wells, and agricultural wells. Many of monitoring studies were conducted in areas that had high pesticide use and agricultural production. Approximately 2% of the ground water samples collected through this program, from 1992 to 1996, had positive detections of diazinon. However, the maximum concentration value was below the limit of quantitation for all wells sampled, and the median value was non-detectable (ND) or <0.002 μg/L. Results from the NAWQA database indicate that diazinon was detected more frequently in shallow ground water in urban areas than in agricultural areas. The concentrations of diazinon in ground water (all wells) ranged from ND to 1.0 μg/L.

4.2.3 Surface Water

After diazinon's residential use started phasing out in June 2003, and after its retail sales ended in December 2004, the concentrations found in urban watersheds reportedly declined nationwide (Banks et al. 2005; Embrey and Moran 2006; US ATSDR 2008; US EPA 2007; USGS 2007). Similarly, NAWQA monitoring data in California urban watersheds from 1995 to 2005 appeared to show steady declines in annual maximum and average concentrations since 2003 (Table 4). Although the frequency of detection fluctuated during this time period, the annual maximum concentration in 2005 and average concentrations in 2004 and 2005 were below diazinon's chronic water criterion of 0.1 μg/L that was developed by CVRWQCB (2006) for California Central Valley and San Joaquin Estuary. This criterion is more stringent than the water criteria and aquatic life benchmarks proposed by the US EPA (2006, 2011) (Table 3). Although concentrations have decreased in California urban waters, diazinon is frequently detected in California's agricultural watersheds at concentrations

Table 4 Diazinon concentrations detected in California urban surface watersheds from 1995 to 2005. Data sources are from USGS National Water Quality Assessment

Year	Number of detections	Percentage detection	Maximum concentration (µg/L)	Mean concentration (µg/L)	Standard deviation	90th percentile concentration
1995	16	100	1.100	0.640	0.274	0.915
1996	2	100	0.337	0.277	0.086	0.325
1997	24	100	1.380	0.370	0.285	0.673
1998	5	83.3	0.420	0.223	0.150	0.362
1999	11	52.4	0.198	0.044	0.060	0.095
2000	26	74.3	0.774	0.106	0.204	0.336
2001	29	93.5	0.947	0.342	0.326	0.773
2002	15	88.2	0.430	0.141	0.142	0.347
2003	13	81.3	0.588	0.147	0.164	0.306
2004	15	78.9	0.218	0.056	0.059	0.128
2005	18	100	0.085	0.035	0.021	0.068

Table 5 A summary of statistical results for diazinon monitoring data in California agricultural counties (2005–2010)[a]

Region	Number of sites	Number of samples	Maximum concentration (µg/L)	Percentage of detection	Number of exceed.	Percentage of exceed.
Sacramento	73	850	2.5	30.2	44	5.2
San Joaquin	121	2,465	1.2	10.0	60	2.4
Salinas	33	244	24.5	91.0	151	61.9
Santa Maria	12	21	0.98	90.5	14	66.7
Imperial	12	58	3.24	51.7	14	24.1

[a]A chronic water quality criterion of 0.1 µg/L was used to determine number of exceedances and percent exceedance

exceeding the water quality criterion of 0.1 µg/L (Starner 2009; Zhang and Starner 2011) (Table 5). Sacramento, Salinas, San Joaquin, Santa Maria, and Imperial counties in California are the top five agricultural regions that have the highest diazinon use, highest detection frequencies, and percentages of exceedance to the chronic water criterion from 2005 to 2010 (Table 5). The range of detection frequencies in these areas was 10–91%. The percentages of exceedance to the water quality criterion ranged from 2.4 to 66.7%. Salinas was the county with the highest detection frequency and percentage of samples exceeding the water quality criterion, followed by Santa Maria, Imperil, Sacramento, and San Joaquin counties (Table 5) (Zhang and Starner 2011). The Salinas River watershed receives both urban and agricultural runoff in the region. Hunt et al. (2003) reported significant acute toxicity to *Ceriodaphnia dubia* in 11% of the main river samples, 87% of the samples from a channel receiving urban and agricultural runoff, 13% of the samples from channels draining agricultural tile drain runoff, and 100% of the samples from a channel draining agricultural surface furrow runoff. Toxicity identification evaluations conducted in 12 samples implicated that the organophosphate pesticides diazinon and chlorpyrifos were probable causes of the observed toxicity in 2/3 of the samples. Chemical analyses also confirmed that sufficient diazinon and/or chlorpyriphos concentrations were detected in samples that caused at least 50% *C. dubia* mortality ($n=31$).

Diazinon can persist in aquatic environments for as long as 6 months. In water, diazinon is subject to both abiotic degradation (i.e., hydrolysis and photolysis) and biotic degradation by microorganisms. The rates of both processes are strongly influenced by pH, temperature, salinity, and the organic content of the water (Larkin and Tjeerdema 2000; US ATSDR 2008). Among the several abiotic transformations that diazinon might undergo in natural waters, hydrolysis and redox reactions are the most common (Brooke and Smith 2005; Wolfe et al. 1990). Frank et al. (1991) observed that temperature had a greater effect on degradation and suggested that hydrolysis was the primary mode of diazinon degradation. Similarly, Garcia-Repetto et al. (1994) and Bondarenko et al. (2004) reported higher diazinon degradation rates under high temperature and more acidic conditions, and reached a similar conclusion. The major diazinon hydrolysis products were 2-isopropyl-4methyl-6-hydroxyprimidine and diethyl thiophosphoric acid or diethyl phosphoric acid (Larkin and Tjeerdema 2000; US EPA 2004). Diazinon undergoes aqueous base-catalyzed and acid-catalyzed hydrolysis and displays the longest hydrolysis half-life near a neutral pH. Morgan (1976) reported a diazinon hydrolysis half-life value as 43.3 days in well water that was between pH 7.4 and 7.7 at 16°C. Rapid diazinon degradation occurred under both higher acidic and basic conditions (Bondarenko et al. 2004; Chapman and Cole 1982; Garcia-Repetto et al. 1994; Frank et al. 1991). Chapman and Cole (1982) reported that pH was the most influential factor on the half-life of diazinon maintained in sterile water–ethanol (99:1) phosphate buffer solutions at 25 ± 3°C. Degradation of diazinon was most rapid under acidic conditions, with half-life values of 3.15, 14, 54.6, 70, and 53.9 days at pH values of 4.5, 5.0, 6.0, 7.0, and 8.0, respectively. There was a similar pH effect on diazinon degradation in work reported by Garcia-Repetto et al. (1994). Bondarenko et al. (2004) investigated the persistence of diazinon in natural waters of various salinities collected from different locations within the Upper Newport Bay-San Diego Creek watershed area located in central Orange County, California. They observed that diazinon degraded fastest in natural water (half-life of 6.3–14.0 days), followed by sea water (half-life of 41.0 days), whereas in sterilized water, diazinon had a half-life of 51.1–54.9 days at 21°C. When the temperature was lowered to 10°C, diazinon persisted much longer, with half-life values of 25.0–28.3 days in natural water; the corresponding half-life value for seawater was 124.0 days. These results suggest that, under similar pH and temperature conditions, diazinon persists the longest in seawater, primarily from lack of microbial degradation in high-saline waters. Diazinon was degraded primarily by abiotic processes in seawater, and any reduction in microbial degradation processes would be expected to produce prolonged persistence. Results also show that sterilization greatly increased diazinon's persistence in freshwater, indicating that as for seawater microbial activity was mainly responsible for its degradation. A comparative study on the degradation of diazinon in sterilized and nonsterilized natural or distilled waters was performed; degradation was more rapid in natural water (12 weeks) than in any other sterilized or distilled waters (>16 weeks), suggesting that microbial degradation occurred in the natural water (Sharom et al. 1980b).

Photolysis is relatively unimportant for degrading diazinon in aquatic systems (US EPA 1976; US ATSDR 2008). In surface water and groundwater samples, diazinon was degraded at similar rates in the light and dark (0.8 vs. 0.7% loss per day) (Frank et al. 1991). Medina et al. (1999) compared the half-life of diazinon in filtered natural water samples held under light and dark conditions and found that sunlight-exposed samples had a half-life of 31.13 days, only 6 days shorter than samples held in the dark. Frank et al. (1991) and Medina et al. (1999) conducted photolysis studies on diazinon and concluded that the contribution of photolysis to degradation was minor.

Acting either independently or together, temperature, pH, salinity, and microbial content are major factors that affect the persistence of diazinon.

4.3 Air and Precipitation

Diazinon released to the atmosphere is subject to direct photolysis, since it adsorbs light primarily in the ultraviolet (UV) region (10–400 nm) (Bavcon et al. 2003; Feigenbrugel et al. 2005). Muñoz et al. (2011) estimated the vapor phase half-life of diazinon in the troposphere as being approximately 4 h as a result of its reaction with photochemically produced hydroxyl radicals. However, in their experiment at 28°C, Muñoz et al. (2011) observed diazinon's half-life, with respect to direct photolysis, to be greater than 1 day. This indicates that the tropospheric degradation of diazinon is dominantly controlled by reaction with hydroxyl radicals rather than by direct photolysis. The pathway for degradation of diazinon in air is depicted in Fig. 7. Diazinon may be transported to the atmosphere from agricultural air-blast spraying or from post-application volatilization (Glotfelty et al. 1990a; McKinney 2005).

One degradation process for diazinon is photochemical oxidation in air. This process, known as oxidative desulfuration, involves addition of OH radicals to P=S bond, resulting in a P=O bond and eventually diazoxon, which is more toxic than diazinon (Glotfelty et al. 1990a; Raina et al. 2010; US EPA 1999; Zhou et al. 2011). Diazoxon is itself hydrolyzed into IMP (2-isopropyl-6-methyl-4-Pyrimidinol) and diethylphosphate by reactions catalyzed by H^+ ions. Another diazinon degradation process involves direct hydrolysis to form products of IMP (2-isopropyl-6-methyl-4-Pyrimidinol) and diethylthiphosphate (Bavcon et al. 2003; McKinney 2005).

When diazinon partitions into droplets in accordance with Henry's Law, it may be hydrolyzed in water vapor formed from fog or rain (McKinney 2005). McKinney (2005) indicated that diazinon may be incorporated in fog and rain drops when they collide with diazinon-containing aerosol particles. Since atmospheric diazinon vapor has a longer equilibration time with fog than rain, due to much longer residence time in fog, it is possible that greater concentrations of diazinon vapor will be incorporated into fog.

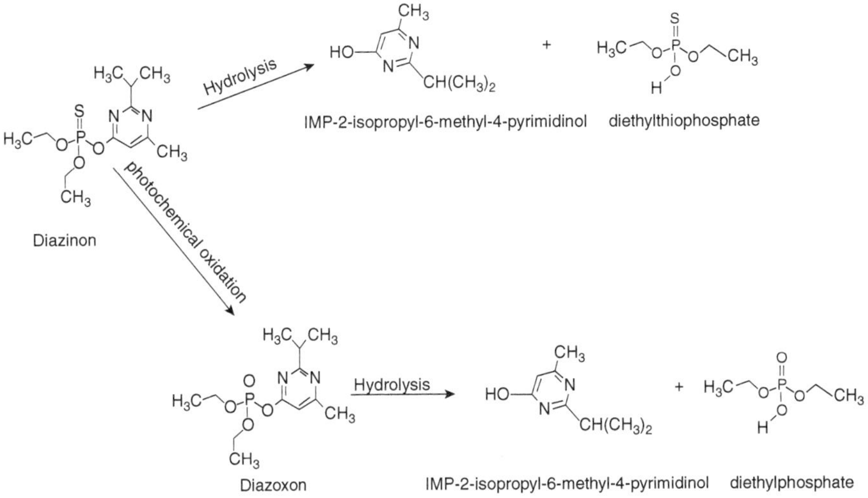

Fig. 7 Diazinon degradation pathways in air

Glotfelty et al. (1990b) studied distribution, drift, and volatilization losses of diazinon to the atmosphere during spray application to a dormant peach orchard. They concluded that distribution and drift losses contributed less diazinon to the atmosphere than the long-term volatilization loss from the soil surface of California's Central Valley. Zabik and Seiber (1993) determined that the organophosphate pesticides, including diazinon, are transported from California's Central Valley to the Sierra Nevada Mountains via several atmospheric processes such as direct air movement, and wet deposition with snow and rain. They observed that pesticide concentrations decreased with distance and elevation from the application site. Diazinon concentrations, which ranged from 13 to 13,000 pg/m^3 at the 114-m elevation, were at times more than a factor of 1,000 greater than those determined (from 1.4 to 83 pg/m^3) at the 533-m elevation. Further, they observed that the concentrations of all compounds appeared to peak during the first 2 weeks of February, within days to weeks after dormant spray applications. At the 1,920-m elevation, the air concentration of diazinon was lower than quantification level (0.7 pg/m^3), whereas the concentration of diazinon in wet-deposition samples was detected at levels up to 48 pg/mL (Zabik and Seiber 1993). McConnell et al. (1998) conducted an extended investigation of atmospheric inputs of pesticides transported from California's Central Valley to the Sierra Nevada Mountains, by collecting winter–spring precipitation (rain and snow) samples from Sequoia National Park (SNP) and from the Lake Tahoe basin. Their study showed that the pesticides currently used in California's Central Valley, including diazinon, were detected in snow and rain samples from two elevations (533 and 1,920 m) in Sequoia National Park in the southern Sierras at much higher levels (<0.21–19 ng/L at 533 m, and <0.057–14 ng/L at 1,920 m) than were samples collected in the Zabik and Seiber (1993) study. The diazinon concentration in Lake Tahoe basin at 2,200-m elevation was found in the range of 0.057–7 ng/L. This indicated that the Lake Tahoe basin snow generally had lower concentrations than those from SNP. This difference in concentration reflects the proximity of SNP to downwind pesticide usage vs. the Lake Tahoe basin (McConnell et al. 1998).

4.3.1 Air Monitoring in California

The extensive use of pesticides on crops could result in air contamination over a wide area for several weeks because of pesticide volatility and drift from application sites. Sava (1985) conducted a 3-day study (samples collected for 6-h period) in residential areas of Salinas, Monterey County, to determine the level of diazinon in ambient air. Eight samples collected during the period from midnight until pesticides were applied to nearby fields the next day gave positive results (Sava 1985). However, with one exception no diazinon was detected between 12:00 p.m. and 12:00 a.m.; this exception was due to reduced application activity and higher wind speeds.

Seiber et al. (1993) conducted an ambient air monitoring study to assess the airborne concentrations of diazinon insecticide, when used as dormant sprays on deciduous fruit and nut orchards. For 24-h ambient time-weighted air samples, the maximum diazinon concentration was 0.307 $\mu g/m^3$ after diazinon was sprayed in orchards 1–100 km from the samplers. Studies conducted in Fresno and Monterey counties for determining aerial movement and deposition of diazinon resulted in the maximum detected concentration of 0.0357 $\mu g/m^3$ and 0.0004 $\mu g/m^3$, respectively (Stein and White 1993). Air Resource Board (ARB) (1998a) conducted an ambient air monitoring study to coincide with the use of diazinon as an insecticide on dormant orchards at five different locations in Fresno County during winter of 1997. Of the 121 ambient samples taken, 21.5% were found to exceed the limit of quantitation (LOQ), 47.1% were found to be at a level between the limit of detection (LOD) and LOQ, and 31.4% were found to be below the LOD. The highest concentration reported was 290 ng/m^3. Segawa et al. (2003) conducted ambient air monitoring for 31 pesticides and their breakdown products in Santa Barbara County, California. They reported maximum diazinon concentrations for the 24 h, 14 day, and 10 week periods as being 0.0021, 0.00087, and 0.00054 $\mu g/m^3$, respectively. Recently, Wofford et al. (2009) conducted a study to determine whether residents of Parlier, Fresno County, were exposed to pesticides in air by establishing three air-sampling locations. The highest 1-day and highest 14-day average concentration was 0.172 $\mu g/m^3$, which is above the health screening level of 0.130 $\mu g/m^3$ and 0.0204 $\mu g/m^3$, respectively. Recently, ARB conducted an ambient air monitoring study in Monterey, San Benito, and Santa Clara counties of California, to determine whether the level of air emissions of diazinon posed a present or potential exposure hazard (Rider 2010a). They reported that approximately 84% of the 192 ambient samples had concentrations less than the method detection limit (MDL) of 0.00154 $\mu g/m^3$. The detectable levels found in the study were between 0.00281 and 0.0173 $\mu g/m^3$.

In addition to the studies cited above, several researchers measured short-term acute air concentrations from individual applications. For example, the Air Resource Board (ARB) (1998b) conducted a diazinon monitoring study on a 40 A dormant peach orchard in Kings County, California, from January 12 to February 2, 1998. They collected 28 samples and found all samples to exceed the Estimated Quantitative Level (EQL) of 44.5 ng/sample, which would be equivalent to 10 ng/m^3 for 24-h sampling period at 3 L/min. The highest diazinon concentration was 5,500 ng/m^3 and was measured during the third (4 h) sampling period (ARB 1998b). In another diazinon air monitoring study in Glenn County, California, Rider (2010b) reported concentrations ranging from less than the MDL to maximum of 4,261 ng/m^3.

Results from the studies conducted in several areas of California indicate that although diazinon is found in ambient air the concentrations of diazinon observed do not warrant immediate regulatory action. However, more research is needed to assess health risks of residue levels found in environmental samples and to determine if exposures should be reduced.

5 Aquatic Toxicology

5.1 Mode of Action

Diazinon shares a common mechanism of toxicity with other organophosphorus insecticides. It inhibits the enzyme acetylcholinesterase, which hydrolyzes the neurotransmitter acetylcholine in cholinergic synapses and neuromuscular junctions. Inhibition of acetylcholinesterase results in accumulation of acetylcholine at neuron synapses causing prolonged simulation of cholinergic receptors. As a consequence, the prolonged neuron stimulation leads to a suite of intermediate syndromes including anorexia, diarrhea, generalized weakness, muscle tremors, abnormal posturing and behavior, depression, and death (Larkin and Tjeerdema 2000; US EPA 2007).

Diazinon undergoes oxidative desulfurization to form a much more toxic degradate, diazoxon. Diazinon belongs to a chemical subgroup of the organophosphorous pesticides called the organophosphorothiolates. Members of this group are poor inhibitors of cholinesterases. Once absorbed into organisms (target and non-target) the thioates are converted to an oxon form by various cytochrome P450 enzymes. Because animals metabolize the parent compound to the oxon, they are exposed both to the parent thioate residue and the highly toxic oxon form (US EPA 2007).

5.1.1 Aquatic Toxicity

The effects that diazinon has on aquatic life have been extensively tested in numerous species under various conditions. Approximately 250 original toxicological studies that include open literature, testing reports from government agencies and registrants were identified by Palumbo et al. (2010). The aquatic toxicity and environmental fate of diazinon has been comprehensively reviewed by several authors, including Larkin and Tjeerdema (2000), US EPA Office of Prevention, Pesticides and Toxic Substances (2004), US EPA Office of Pesticide Programs (2007), and US ATSDR (2008). The reviewers had different objectives and used different data evaluation criteria in their individual reviews. Moreover, they relied on different data sets from nonidentical sources, and the result is that discrepancies exist in the range of toxicity values they reported. However, the differences are not significant enough to alter the toxicity ratings for diazinon given for various aquatic taxa. In this review, the toxicity data sets evaluated by Palumbo et al. (2010) in freshwater fish and invertebrates were used to summarize diazinon's toxicity, because the method used by these authors provided an evaluation of toxicity that was more stringent and consistent.

5.1.2 Acute Toxicity

Diazinon toxicity varies widely within and among species, and varies when performed under different testing conditions. Reliable acute toxicity values for diazinon were identified for four fish species and nine invertebrate species (Table 6). The 96-h LC_{50} values reported ranged from 0.21 µg/L for *C. dubia* to 10,000 µg/L for fathead minnow, *Pimephales promelas*; this level of toxicity was rated by EPA (US EPA 2004) as being very highly to moderately toxic to aquatic organisms. In general, freshwater cladocerans are more sensitive than freshwater teleosts and aquatic plants. The most sensitive organisms tested were members of the family Daphniidae. *C. dubia* had the lowest species mean acute value (SMAV) of 0.36 µg/L, and a similar acute value of 0.52 µg/L was reported for the Daphniidae, *Daphnia magna*. Two amphipods (*Hyalella azteca* and *Gamarus pseudolinaeus*) were tested and had 96-h LC_{50} values ranging from 4.3 to 16.8 µg/L. Two insects (*Procloeon* sp., *Chironomus tentans*) and one mysid (*Neomysis mercedis*) were tested, and the resulting 96-h LC_{50} SMAVs ranged from 1.79 to 10.7 µg/L, which is in the range similar as the amphipods. Snails (*Physa* spp. and *Pomacea paludosa*) were the invertebrates that had the lowest sensitivity to diazinon. The SMAVs for snails ranged from 3,198 to 4,441 µg/L. Diazinon is highly or moderately toxic to fish species and displays 96-h LC_{50}s for them ranging from 440 to 10,000 µg/L. The bluegill sunfish *Lepomis acrochirus* was the most sensitive fish species tested, and had an SMAV of 460 µg/L. A similar SMAV (723 µg/L) was reported for brook trout *Salvelinus fontinalis*. Flagfish *Jordanella floridae* and fathead minnow were moderately sensitive to diazinon and had SMAVs of 1,643 µg/L and 6,900–9,250 µg/L, respectively. The only available aquatic plant study was performed with green algae (*Pseudokirchneriella subcapitata*) and produced a 7-day EC_{50} value of 3,700 µg/L (US EPA 2007).

5.1.3 Chronic Toxicity

Two freshwater fish species and one invertebrate species were tested for chronic toxicity (Palumbo et al. 2010; US EPA 2005) (Table 7). The invertebrate, *D. magna*,

Table 6 Acute toxicity of diazinon to freshwater aquatic organisms

Species	Duration hour	Age/size	$LC_{50}/\mu g/L$ $SMAV^a$	References
Flagfish *Jordanella floridae*	96	6–7 weeks	1,643	Allison and Hermanutz (1977)
Bluegill *Lepomis acrochirus*	96	1 year	460	Allison and Hermanutz (1977)
Fathead minnow *Pimephales promelas*	96	15–20 weeks	7,656	Allison and Hermanutz (1977)
Fathead minnow *P. promelas*	96	31 days	9,350	Geiger et al. (1988)
Fathead minnow *P. promelas*	96	Newly hatched	6,900	Jarvinen and Tanner (1982)
Brook trout *Salvelinus fontinalis*	96	1 year	723	Allison and Hermanutz (1977)
Cladoceran *Ceriodaphnia dubia*	96	<24 h	0.36	Bailey et al. (1997, 2000, 2001), Banks et al. (2005), CDFG (1992a, 1998a)
Cladoceran *Daphnia magna*	96	<24 h	0.52	Ankley and Collyard 1995
Amphipod *Gammarus pseudolimnaeus*	96	Mature	16.82	Hall and Anderson (2005)
Amphipod *Hyalella azteca*	96	14–21 days	4.3	Anderson and Lydy (2002)
Insect *Chironomus tentans*	96	3rd instar	10.7	Ankley and Collyard (1995)
Insect *Procloeon* sp.	48	0.5–1 cm	1.79	Anderson et al. (2006)
Mysid *Neomysis mercedis*	96	<5 days	4.15	CDFG (1992b)
Pond snail *Physa* spp.	96	Juvenile	4,441	CDFG (1998b)
Snail *Pomacea paludosa*	96	1 day, 7 days	3,198	Jarvinen and Tanner (1982)

[a]SMAV: species mean acute value

Table 7 Chronic toxicity of diazinon to freshwater aquatic organisms

Species	Duration	Age/size	NOEC[a] (µg/L)	References
Cladoceran *Daphnia magna*.	21 days	<24 h	0.17	Surprenant (1988)
Brook trout *Salvelinus fontinalis*	173 days	1 year	4.8	Allison and Hermanutz (1977)
Fathead minnow *Pimephales promelas*	274 days	5 days	28	Allison and Hermanutz (1977)
Fathead minnow *P. promela*	32 days	Newly hatched	50	Jarvinen and Tanner (1982)

[a]NOEC: on-observed effect concentration

was most sensitive to the chronic toxicity of diazinon in a 21-days life cycle test. The NOEC and LOEC values in this study, based on a survival endpoint, were 0.17 µg/L and 0.32 µg/L, respectively (Surprenant 1988). A 173-days partial life cycle test with brook trout indicated that the survival of parental stock was unaffected at diazinon concentrations below 4.8 µg/L (Allison and Hermanutz 1977). Fathead minnows tested from 5-day post-hatch to spawning showed a significant higher incidence of scoliosis and reduced hatching rates than did controls at concentrations above 3.2 µg/L (Surprenant 1988). However, according to the US EPA (2007), a chronic NOEC in freshwater fish could not be established because data did not meet test guideline requirements. Currently, the lowest concentration tested using brook trout (<0.55 µg/L) is used as the NOEC for diazinon in freshwater fish.

5.2 Bioaccumulation

Diazinon bioaccumulates, but demonstrates a wide range of accumulation rates and efficiencies among different aquatic organisms (Palumbo et al. 2010; US ATSDR 2008). Its bioconcentration factor (BCF) values ranged from 4 for shrimp (Seguchi and Asaka 1981) to 300 for zebrafish (Keizer et al. 1991). In general, freshwater fish had higher bioconcentration ratios than crustaceans and gastropods (Kanazawa 1978). Differences in metabolism among species and exposure concentrations play a role in determining BCF. For example, guppies (LC_{50}=0.8 mg/L) were ten times more sensitive to diazinon than were zebrafish (LC_{50}=8 mg/L). Their BCF values also varied greatly from 39 for guppies to >300 for zebrafish (Keizer et al. 1991). In another study, diazinon BCF values for guppies ranged from 148 to 224, depending upon exposure concentrations (Deneer et al. 1999). Diazinon accumulated quickly in most of the tested fish and its BCF values plateaued within a few days. Japanese killifish, exposed to a mixture of pesticides that included diazinon, reached a BCF plateau within 24 h (Tsuda et al. 1995). The BCF values for four fish species (guppy, killifish, goldfish, and white cloud mountain fish) exposed to 2.1–2.9 µg/L diazinon peaked within 120 h in all fish, and ranged from 37.5 in the white cloud mountain fish to 132 in the male guppy (Tsuda et al. 1997). Diazinon is quickly eliminated from tissues of tested animals within a few days (3–8) of exposure, once they are placed in clean water. Sancho

Table 8 Aquatic-life benchmark values and water quality criteria for diazinon

		Benchmark/water quality criteria	
		Acute (µg/L)	Chronic (µg/L)
Benchmark	Fish	45	<0.55
	Invertebrate	0.11	0.17
	Nonvascular plant	6,700	N/A
	Vascular Plant	N/A	N/A
Water quality criteria	US EPA[a]	0.17	0.17
	CVRWQCB[b]	0.16	0.1
	Palumbo et al. (2010)	0.2	0.07

[a]US EPA: US Environmental Protection Agency
[b]CVRWQCB: Central Valley Regional Water Resource Quality Control Board

et al. (1992) observed that >50% diazinon in a freshwater eel was eliminated after 24 h in clean water. El Arab et al. (1990) determined that only 9% of bioaccumulated diazinon was left in perch after 3 days. Kanazawa (1978) treated seven freshwater species, including fish and crustaceans, to 10 and 50 µg/L levels of diazinon for 7 days and found that diazinon was eliminated from tissues within 8 days.

5.3 Aquatic Life Benchmarks and Water Quality Criteria

Aquatic life benchmark and water quality criteria are toxicological reference values developed separately by EPA's Office of Pesticide Program (OPP) and Office of Water (OW) that use different approaches (US EPA 2011). Both EPA offices have responsibilities for evaluating aquatic toxicity data to assess the ecological effects of chemicals in surface water. OPP develops the aquatic life benchmarks for freshwater species based on toxicity values. Each Aquatic Life Benchmark is based on the most sensitive, scientifically acceptable toxicity endpoint for a given taxon (e.g., freshwater fish). OPP uses aquatic toxicity data in ecological risk assessments for large numbers of pesticide registration decisions under the Federal Insecticide, Fungicide, and Rodenticide Act (FIFRA). The diazinon aquatic life benchmarks are available for each taxon and these are shown in Table 8. OW uses aquatic toxicity data to develop ambient water quality criteria that can be adopted by states and Indian tribes on reservation land to establish water quality standards under the Clean Water Act. Based on the procedures described in the "Guidelines for Deriving Numerical National Water Quality Criteria for the Protection of Aquatic Organisms and Their Uses" (Stephan et al. 1985), diazinon water quality criteria were 0.17 µg/L for both of the criteria: continuous concentration and criterion maximum concentration (Table 8) (US EPA 2005, 2011). The California Central Valley Regional Water Resources Control Board (CVRWQCB 2006) developed its own water quality criteria for the control of diazinon runoff into the Sacramento-San Joaquin Delta by using US EPA's methodology, but with slightly different data set. This resulted in an acute water quality criterion for diazinon of 0.16 µg/L and a chronic water criterion of 0.1 µg/L, which are more stringent criteria than those of the US EPA. Palumbo et al. (2010) evaluated all of the existing methodologies worldwide and reported a detailed approach for water quality criteria derivation. Based on their methodology, they recommended diazinon water quality criteria of 0.2 and 0.07 µg/L for acute and chronic exposures, respectively.

6 Summary

Diazinon, first introduced in USA in 1956, is a broad-spectrum contact organophosphate pesticide that has been used as an insecticide, acaricide, and nematicide. It has been one of the most widely used insecticides in the USA for household and agricultural pest control. In 2004, residential use of diazinon was discontinued, and in 2009, diazinon was phased out of all agricultural uses. Consequently, the amounts of diazinon applied have been drastically decreased. For example, in California, the amount of diazinon applied decreased from 501,784 kg in 2000 to 64,122 kg in 2010.

Diazinon has a K_{oc} value of 40–432, and is considered to be moderately mobile in soils. Diazinon residues have been detected in groundwater, drinking water wells, monitoring wells, and agricultural wells. The highest detection frequencies and highest percentages of exceedance of the water quality criterion value of 0.1 µg/L have been reported from the top five agricultural counties in California that had the highest diazinon use. Diazinon is transported in air via atmospheric processes such as direct air movement and wet deposition in snow and rain, although concentrations decrease with distance and elevation from the source. In the environment, diazinon undergoes degradation by several processes, the most important of which is microbial degradation in soils. The rate of diazinon degradation is affected by pH, soil type, organic amendments, soil moisture, and the concentration of diazinon in the soil, with soil pH being a major influencing factor in diazinon degradation rate. Studies indicate that soil organic matter is the most important factor that influences diazinon sorption by soils, although clay content and soil pH also play an important role in diazinon sorption.

Diazinon is very highly to moderately toxic to aquatic organisms. Diazinon inhibits the enzyme acetylcholinesterase, which hydrolyzes the neurotransmitter acetylcholine and leads to a suite of intermediate syndromes including anorexia, diarrhea, generalized weakness, muscle tremors, abnormal posturing and behavior, depression, and death. Differences in metabolism among species and exposure concentrations play a vital role in diazinon's bioaccumulation among different aquatic organisms in a wide range of accumulating rates and efficiencies.

Acknowledgements Support was provided by the Environmental Monitoring Branch of the Department of Pesticide Regulation (CDPR), California Environmental Protection Agency. The statements and conclusions are those of the authors and not necessarily those of CDPR. The mention of commercial products, their sources, or their use in connection with materials reported herein is not to be construed as actual or implied endorsement of such products.

References

Allison DT, Hermanutz RO (1977) Toxicity of diazinon to brook trout and fathead minnows. EPA-600/3-77-060. Environmental Research Laboratory-Duluth, Office of Research and Development, US Environmental Protection Agency, Duluth, MN

Anderson TD, Lydy MJ (2002) Increased toxicity to invertebrates associated with a mixture of atrazine and organophosphate insecticides. Environ Toxicol Chem 21:1507–1514

Anderson BS, Phillips BM, Hunt JW, Connor V, Richard N, Tjeerdema RS (2006) Identifying primary stressors impacting macroinvertebrates in the Salinas River (California, USA): Relative effects of pesticides and suspended particles. Environ Poll 141:402–408

Ando C, Leyva J, Gana C (1993) Monitoring diazinon in the Mediterranean fruit fly eradication soil treatment program. California Department of Pesticide Regulation, Los Angeles, CA

Ankley GT, Collyard SA (1995) Influence of piperonyl butoxide on the toxicity of organophosphate insecticides to three species of freshwater benthic invertebrates. Comp Biochem Physiol 110C:149–155

Air Resource Board (1998a) Report for the ambient air monitoring of diazinon in Fresno County during winter, 1997. State of California. California Environmental Protection Agency, Air Resources Board, Sacramento, CA. doi:www.cdpr.ca.gov/docs/emon/pubs/tac/tacpdfs/diaamb.pdf

Air Resource Board (1998b) Report for the application (Kings County) and ambient (Fresno County) air monitoring of diazinon during winter, 1998. State of California, California Environmental Protection Agency, Air Resources Board, Sacramento, CA, http://www.cdpr.ca.gov/docs/emon/pubs/tac/tacpdfs/diamapl.pdf

Arienzo M, Sanchezcamazano M, Herrero TC, Sanchezmartin MJ (1993) Effect of organic cosolvents on adsorption of organophosphorus pesticides by soils. Chemosphere 27(8):1409–1417

Arienzo M, Herrero TC, Sanchezmartin MJ, Sanchezcamazano M (1994a) Effect of Soil Characteristics on Adsorption and Mobility of (C-14)Diazinon. J Agric Food Chem 42(8):1803–1808

Arienzo M, Sanchezcamazano M, Sanchezmartin MJ, Herrero TC (1994b) Influence of Exogenous Organic-Matter in the Mobility of Diazinon in Soils. Chemosphere 29(6):1245–1252

Armstrong DE, Chesters G, Harris RF (1967) Atrazine hydrolysis in soil. Soil Sci Soc Am Proc 31:61–66

Bailey HC, Miller JL, Miller MJ, Wiborg LC, Deanovic L, Shed T (1997) Joint acute toxicity of diazinon and chlorpyrifos to *Ceriodaphnia dubia*. Environ Toxicol Chem 16:2304–2308

Bailey HC, Draloi R, Elphick JR, Mulhall A-M, Hunt P, Tedmanson L, Lovell A (2000) Application of *Ceriodaphnia dubia* for whole effluent toxicity tests in the Hawkesbury-Nepean watershed, New South Wales, Australia: method development and validation. Environ Toxicol Chem 19:88–93

Bailey HC, Elphick JR, Drassoi R, Lovell A (2001) Joint acute toxicity of diazinon and ammonia to *Ceriodaphnia dubia*. Environ Toxicol Chem 20:2877–2882

Banks KE, Hunter DH, Wachal DJ (2005) Diazinon in surface waters before and after a federally-mandated ban. Science of the Total Environment 350:86–93

Bavcon M, Trebse P, Zupancic-Kralj L (2003) Investigations of the determination and transformations of diazinon and malathion under environmental conditions using gas chromatography coupled with a flame ionisation detector. Chemosphere 50:595–601

Bondarenko S, Gan J, Haver DL, Kabashima JN (2004) Persistence of selected organophosphate and carbamate insecticides in waters from a coastal watershed. Environ Toxicol Chem 23(11):2649–2654

Brooke LT, Smith GJ (2005) Aquatic life ambient water quality criteria: diazinon. United States Environmental Protection Agency. Office of Water, Office of Science and Technology, Washington, DC, EPA-822-R-05-006

Budavari S, O'Neal MJ, Smith A, Heckelman PE (1989) The Merck index: an encyclopedia of chemicals, drugs and biologicals. Merck and Co., Inc., Rahway, NJ

Burkhard N, Guth JA (1979) Photolysis of organophosphorus insecticides on soil surfaces. Pesti Sci 10:313–319

CALPIP (2011) California pesticide information portal. California Environmental Protection Agency, Department of Pesticide Regulation, Sacramento, CA. http://calpip.cdpr.ca.gov/main.cfm. Accessed 19 Jun 2011

CDFG (California Department of Fish and Game) (1992a) Test No. 162. 96-h acute toxicity of diazinon to *Neomysis mercedis*. Aquatic Toxicity Laboratory, California Department of Fish and Game, Elk Grove, CA

CDFG (California Department of Fish and Game) (1992b) Test No. 168. 96-h acute toxicity of diazinon to *Neomysis mercedis*. Aquatic Toxicology Laboratory, California Department of Fish and Game, Elk Grove, CA

CDFG (California Department of Fish and Game) (1998a) Test No. 122. 96-h acute toxicity of diazinon to *Ceriodaphnia dubia*. Aquatic Toxicology Laboratory, California Department of Fish and Game, Elk Grove, CA

CDFG (California Department of Fish and Game) (1998b) Test 132. 96-h toxicity of diazinon to *Physa* sp. Aquatic Toxicology Laboratory, California Department of Fish and Game, Elk Grove, CA

CDPR (California Department of Pesticide Regulations) (2010a) Pesticide chemistry database. California Environmental Protection Agency, Department of Pesticide Regulation, Sacramento, CA

CDPR (California Department of Pesticide Regulations) (2010b) Top 100 lists: the top 100 pesticides used statewide (all sites combined) in 2010. California Environmental Protection Agency, Sacramento, CA. p 2. http://www.cdpr.ca.gov/docs/pur/pur10rep/top_100_ais_lbs10.pdf. Accessed 9 May 2012

CDPR (California Department of Pesticide Regulations) (2011) Groundwater database. California Environmental Protection Agency, Department of Pesticide Regulation, Sacramento, CA

CVRWQCB (2006) Amendments to the water quality control plan for the Sacramento river and san Joaquin river basins for the control of diazinon and chlorpyrifos runoff in the lower San Joaquin river, final staff report. Central Valley Regional Water Quality Control Board, State Water Resources Control Board, California Environmental Protection Agency, Rancho Cordova, CA

Calvet R, Terce M, Arvieu JC (1980) Adsorption of pesticides by soils and their constituents 3. General-characteristics of the pesticides adsorption. Annales Agronomiques 31(3):239–257

Carey AE, Kutz FW (1985) Trends in ambient concentrations of agrochemicals in humans and the environment of the USA. Environmental Monitoring and Assessment 5:155–164

Chapman RA, Cole CM (1982) Observations on the influence of water and soil pH on the persistence of insecticides. Journal of Environmental Science and Health Part B Pesticides Food Contaminants and Agricultural Wastes 17(5):487–504

Cohen DB (1986) Ground water contamination by toxic substances. A California assessment. In: Garner WY, Honeycutt RC, Nigg HN (eds) Evaluation of pesticides in ground water. American Chemical Society, Washington, DC, pp 499–529

Deneer JW, Budde BJ, Weijers A (1999) Variations in the lethal body burdens of organophosphorus compounds in the guppy. Chemosphere 38:1671–1683

Eitzer BD, Chevalier A (1999) Landscape care pesticide residues in residential drinking water wells. Bulletin of Environmental Contamination and Toxicology 62:420–427

El Arab AE, Attar A, Ballhorn L, Freitag D, Korte D (1990) Behavior of diazinon in a perch species. Chemosphere 21:193–199

Embrey S, Moran P (2006) Quality of stream water in the puget sound basin—a decade of study and beyond. http://wa.water.usgs.gov/projects/pugt/data/4-2006-poster.pdf. Accessed 21 Nov 2011

Feigenbrugel V, Loew C, Le Calvé S, Mirabel P (2005) Near-UV molar absorptivities of acetone, alachlor, metolachlor, diazinon and dichlorvos in aqueous solution. J. Photochem. Photobiol. A Chemistry 174:76–81

Felsot AS, Racke KD, Hamilton DJ (2003) Disposal and degradation of pesticide waste. Reviews of Environmental Contamination and Toxicology 177:123–200

Fenlon KA, Jones KC, Semple KT (2007) Development of microbial degradation of cypermethrin and diazinon in organically and conventionally managed soils. Journal of Environmental Monitoring 9(6):510–515

Frank R, Braun HE, Chapman N, Burchat C (1991) Degradation of parent compounds of nine organophosphorus insecticides in Ontario surface and ground waters under controlled conditions. Bulletin of Environmental Contamination and Toxicology 47(3):374–380

Frank R, Braun HE, Clegg BS, Ripley BD, Johnson R (1990) Survey of farm wells for pesticides, Ontario, Canada, 1986 and 1987. Bulletin of Environmental Contamination and Toxicology 44(3):410–419

Frank R, Clegg BS, Ripley BD, Braun HE (1987) Investigations of pesticide contaminations in rural wells, 1979-1984, Ontario, Canada. Archives of Environmental Contamination and Toxicology 16:9–22

Frederick CM Jr, Graeber D, Forney LJ, Reddy CA (1996) The fate of lawn care pesticides during composting. Biocycle 37(3):64–66

Garcia-Repetto R, Martinez D, Repetto M (1994) The influence of pH on the degradation kinetics of some organophosphorous pesticides in aqueous solutions. Veterinary and human toxicology 36(3):202–204

Geiger DL, Call DJ, Brooke LT (1988) Acute toxicities of organic chemicals to fathead minnows (*Pimephales promelas*). Center for Lake Superior Environmental Studies, University of Wisconsin-Superior, Wisconsin, pp 279–280

Getzin LW, Rosefield I (1966) Persistence of Diazinon and Zinophos in Soils. J Econ Ento 59(3):512–516

Glotfelty DE, Majewski MS, Seiber JN (1990a) Distribution of several organophosphorous insecticides and their oxon analogues in a foggy atmosphere. Environ Sci Technol 24:353–357

Glotfelty DE, Schomburg CJ, McChesney MM, Sagebiel JC, Seiber JN (1990b) Studies of the distribution, drift, and volatilization of diazinon resulting from spray application to a dormant peach orchard. Chemosphere 21:1303–1314

Gunner HB (1970) Microbial ecosystem stress induced by an organophosphate insecticide. Mededelingen Faculteit Landbonwwetenschapen Gent 35:581

Gunner HB, Zuckerman B, Walker RW, Miller CW, Deubert KH, Longley RE (1966) Distribution and Persistence of Diazinon Applied to Plant and Soil and Its Influence on Rhizosphere and Soil Microflora. Plant and Soil 25(2):249–264

Gysin H, Margot A (1958) Chemistry and Toxicological Properties of O, O-Diethyl-O-(2-isopropyl-4-methyl-6-pyrimidinyl) phosphorothioate (Diazinon). J Agric Food Chem 6:900–903

Hall LW Jr, Anderson RD (2005) Acute toxicity of diazinon to the amphipod, *Gammarus pseudolimnaeus*: implications for water quality criteria. Bull Environ Contam Toxicol 74:94–99

Holden PW (1986) Pesticides and groundwater quality: issues and problems in four states. National Academy Press, Board on Agriculture National Research Council, Washington, DC

Hunt JW, Anderson BS, Philips PN, Tjeerdema RS, Puckett HM, Stephenson M, Worcester K, De Vlaming V (2003) Ambient toxicity due to chlowpyrifos and diazinon in a central California coastal watershed. Environmental Monitoring Assessment 82:83–112

Ingram CW, Coyne MS, Williams DW (2005) Effects of commercial diazinon and imidacloprid on microbial urease activity in soil and sod. J Environ Qual 34(5):1573–1580

Jarvinen AW, Tanner DK (1982) Toxicity of selected controlled release and corresponding unformulated technical grade pesticides to the fathead minnow *Pimephales promelas*. Environmental Pollution Series a-Ecological and Biological 27:179–195

Kanazawa J (1978) Bioconcentration ratio of diazinon by freshwater fish and snail. Bull Environ Contam Toxicol 20:613–617

Keizer J, D'Agostino G, Vittozzi L (1991) The importance of biotransformation in the toxicity of xenobiotics to fish. I. Toxicity and bioaccumulation of diazinon in guppy (*Poecilia reticulata*) and zebra fish (*Brachydanio rerio*). Aquat Toxicol 21:239–254

Konrad JG, Armstron DE, Chesters G (1967) Soil Degradation of Diazinon a Phosphorothioate Insecticide. Agron J 59(6):591–594

Larkin DJ, Tjeerdema RS (2000) Fate and effects of diazinon. Reviews Environmental Contamination and Toxicology 166:49–82

Leistra M, Tuinstra L, Vanderburg AMM, Crum SJH (1984) Contribution of Leaching of Diazinon, Parathion, Tetrachlorvinphos and Triazophos from Glasshouse Soils to Their Concentrations in Water Courses. Chemosphere 13(3):403–413

Leland JE, Mullins DE, Berry DF (2003) The fate of C-14-diazinon in compost, compost-amended soil, and uptake by earthworms. Journal of Environmental Science and Health Part B-Pesticides Food Contaminants and Agricultural Wastes 38(6):697–712

Leonard RA (1990) Movement of pesticides into surface waters. In: Cheng HH (ed) Pesticides in soil environment, vol 2. SSSA, Madison, WI, pp 303–350

Levanon D, Meisinger JJ, Codling EE, Starr JL (1994) Impact of tillage on microbial activity and the fate of pesticides in the upper soil. Water Air and Soil Pollut 72(1–4):179–189

Levot GW, Lund RD, Black R (2004) Diazinon and diflubenzuron residues in soil following surface disposal of spent sheep dip wash. Australian J Experimental Agric 44(10):975–982

Lichtenstein EP, Schulz KR (1970) Volatilization of Insecticides from various substrates. J Agric Food Chem 18:814–818

Long T (1989) Groundwater contamination in the vicinity of agrichemical mixing and loading facilities. Paper presented at the proceedings of the Illinois agricultural pesticides conference, Cooperative Extension Service of the University of Illinois, Urbana-Champaign

Mackay D, Shiu WY, Ma KC, Lee SC (2006) Handbook of physical—chemical properties and environmental fate for organic chemicals, vol Nitrogen and sulfur containing compounds and pesticides, IV. CRC, Taylor and Francis, New York

Mansour M, Feicht EA, Behechti A, Scheunert I (1997) Experimental approaches to studying the photostability of selected pesticides in water and soil. Chemosphere 35(1–2):39–50

McConnell LL, Lenoir JS, Datta S, Seiber JN (1998) Wet deposition of current-use pesticides in the Sierra Nevada Mountain range, California, USA. Environ Toxicol Chem 17:1908–1916

McKinney MG (2005) Diazinon and diazinon metabolites in fog: Northern Sacramento Valley, California. Master of Science Thesis in Geosciences, California State University, Chico

Medina D, Prieto A, Ettiene G, Buscema I, de Abreu VA (1999) Persistence of organophosphorus pesticide residues in Limon River waters. Bull Environ Contam Toxicol 63:39–44

Morgan HG (1976) Sublethal effects of diazinon in stream invertebrates. Ph.D. dissertation. NIG 2WI. Univ. of Guelph, Ontario, Canada. p. 157

Muñoz A, Person AL, Calvé SL, Mellouki A, Borrás E, Daële V, Vera T (2011) Studies on atmospheric degradation of diazinon in the EUPHORE simulation chamber. Chemosphere 85:724–730

Nemeth-Konda L, Fuleky G, Morovjan G, Csokan P (2002) Sorption behaviour of acetochlor, atrazine, carbendazim, diazinon, imidacloprid and isoproturon on Hungarian agricultural soil. Chemosphere 48(5):545–552

Palumbo AJ, Fojut TL, TenBrook PL, Tjeerdema RS (2010) Water quality criteria report for diazinon. Phase III: Application of the pesticide water quality criteria methodology. Report prepared for the Central Valley Regional Water Quality Control Board, Rancho Cordova, CA

Petruska JA, Mullins DE, Young RW, Collins ERJ (1985) A benchtop system for evaluation of pesticide disposal by composting. Nuclear and Chemical Waste Management 5(3):177–182

Raina R, Hall P, Sun L (2010) Occurrence and Relationship of Organophosphorus Insecticides and Their Degradation Products in the Atmosphere in Western Canada Agricultural Regions. Environ Sci Technol 44:8541–8546

Rider S (2010a) Ambient pesticide air monitoring for diazinon and diazoxon in Monterey, San Benito and Santa Clara counties during July and August of 2009. California Environmental Protection Agency, Air Resources Board, California

Rider S (2010b) Report on air monitoring of an orchard application of diazinon in Glenn County during January 2010. California Environmental Protection Agency, Air Resources Board, California

Ritter WF, Johnson HP, Lovely WG (1973) Diffusion of Atrazine, Propachlor, and Diazinon in a Silt Loam Soil. Weed Sci 21(5):381–384

Ritter WF, Johnson HP, Lovely WG, Molnau M (1974) Atrazine, Propachlor, and Diazinon Residues on Small Agricultural Watersheds - Runoff Losses, Persistence, and Movement. Environ Sci Technol 8(1):38–42

Sanchez-Camazano M, Sanchez-Martin MJ (1984) The Adsorption of Azinphosmethyl by Soils. Agrochimica 28(2–3):148–158

Sanchez-Martin MJ, Sanchez-Camazano M (1991) Adsorption of Chloridazon by Soils and Their Components. Weed Science 39(3):417–422

Sancho E, Ferrando MD, Andreu E, Gamon M (1992) Acute toxicity, uptake and clearance of diazinon by the European eel, *Anguila Anguilla* (L.). J Environ Sci Health 27B:209–221

Sapozhnikova Y, Bawardi O, Schlenk D (2004) Pesticides and PCB's in sediments and fish from the Salton Sea, California, US. Chemosphere 55:797–809

Sarmah AK, Close ME, Mason NWH (2009) Dissipation and sorption of six commonly used pesticides in two contrasting soils of New Zealand. Journal of Environmental Science and Health Part B Pesticides Food Contaminants and Agricultural Wastes 44(4):325–336

Sattar MA (1990) Fate of Organophosphorus Pesticides in Soils. Chemosphere 20(3–4):387–396

Sava RJ (1985) Monterey County residential air monitoring. The California Environmental Protection Agency, Department of Pesticide Regulation, Environmental Monitoring and Pest Management Branch, California, EH85-07

Schoen SR, Winterlin WL (1987) The effects of various soil factors and amendments on the degradation of pesticide mixtures. Journal of Environmental Science and Health Part B Pesticides Food Contaminants and Agricultural Wastes 22(3):347–377

Segawa R, Schreider J, Wofford P (2003) Ambient air monitoring for pesticides. State of California, Environmental Protection Agency, Department of Pesticide Regulation, Lompoc, CA, EH03-02

Seguchi K, Asaka S (1981) Intake and excretion of diazinon in freshwater fishes. Bull Environ Contam Toxicol 27:244–249

Seiber JN, Wilson BW, McChesney MM (1993) Air and fog deposition residues of four organophosphate insecticides used on dormant orchards in the San Joaquin Valley, California. Environ Sci Technol 27:2236–2243

Sethunathan N, Macrae IC (1969a) Persistence and Biodegradation of Diazinon in Submerged Soils. J Agric Food Chem 17(2):221–225

Sethunathan N, Macrae IC (1969b) Some Effects of Diazinon on Microflora of Submerged Soils. Plant and Soil 30(1):109–112

Sethunathan N, Yoshida T (1969) Fate of Diazinon in Submerged Soil - Accumulation of Hydrolysis Product. J Agric Food Chem 17(6):1192–1195

Sharom MS, Miles JRW, Harris CR, McEwen FL (1980a) Behavior of 12 insecticides in soil and aqueous suspensions of soil and sediment. Water Res 14(8):1095–1100

Sharom MS, Miles JRW, Harris CR, McEwen FL (1980b) Persistence of 12 insecticides in water. Water Res 14(8):1089–1093

Singh J, Singh DK (2005) Bacterial, azotobacter, actinomycetes, and fungal population in soil after diazinon, imidacloprid, and lindane treatments in groundnut (Arachis hypogaea L.) fields. Journal of Environmental Science and Health Part B-Pesticides Food Contaminants and. Agricultural Wastes 40(5):785–800

Starner K (2009) Spatial and temporal analysis of diazinon irrigation-season use and monitoring data. Report. California Department of Pesticide Regulation, Sacramento, CA. Available via CDPR: http://www.cdpr.ca.gov/docs/emon/surfwtr/policies/starner_sw08.pdf. Accessed 15 Aug 2011

Stein RG, White JH (1993) Aerial movement and deposition of diazinon, chlorpyrifos, and ethyl parathion. State of California, Environmental Protection Agency, Department of Pesticide Regulation, Environmental Monitoring and Pest Management Branch, California, EH 93-04

Stephan CE, Mount DI, Hanson DJ, Gentile JH, Chapman GA, Brungs WA (1985) Guidelines for deriving numerical national water quality criteria for the protection of aquatic organisms and their uses. EPA PB85-227049

Surprenant DC (1988) The chronic toxicity of 14C-diazinon technical to *Daphnia magna* under flow-through conditions, EPA guidelines No. 72-4. Agricultural Division, Ciba-Geigy Corporation, Greensboro, NC

Szeto SY, Wan MT, Price P, Roland J (1990) Distribution and persistence of diazinon in a cranberry bog. J Agric Food Chem 38(1):281–285

The International Union of Pure and Applied Chemistry (IUPAC) (2010) Agrochemicals in footprint database. Available online from http://sitem.herts.ac.uk/aeru/iupac/index.htm. Accessed 12 Aug 2010

Tsuda T, Aoki S, Inoue T, Kojima M (1995) Accumulation and excretion of diazinon, fenthion and fenitrothion by killifish: Comparison of individual and mixed pesticides. Water Res 29:455–458

Tsuda T, Kojima M, Harada A, Nakajima AS (1997) Relationships of bioconcentration factors of organophosphate pesticides among species of fish. Comp Biochem Physiol 116C:213–218

US Agency for Toxic Substances and Disease Registry (ATSDR) (2008) Toxicological profile for Diazinon. Department of Health and Human Services, Atlanta, GA

US Geological Survey (USGS) (2007) National water quality assessment program. http://water.usgs.gov/nawqa/. Accessed 3 May 2007

US EPA (1976) Chemical and photochemical transformation of selected pesticides in aquatic systems. US Environmental Protection Agency, Athens, GA, EPA600376067

US EPA (1999) Environmental risk assessment for diazinon. Office of Prevention, Pesticides and Toxic Substances, Washington, DC

US EPA (2000) EPA announces elimination of all indoor uses of widely used pesticide Diazinon; begins phase-out of lawn and garden uses [Online]. Available at http://yosemite.epa.gov/opa/admpress.nsf/a883dc3da7094f97852572a00065d7d8/c8cdc9ea7d5ff585852569ac0077bd31!OpenDocument. Accessed 23 Jan 2011

US EPA (2004) Interim reregistration eligibility decision for diazinon. United States Environmental Protection Agency, Office of Prevention, Pesticides and Toxic Substances, Washington, DC, EPA 738-R-04-066

US EPA (2005) Aquatic life ambient water quality criteria-diazinon. EPA Office of Water, Washington, DC, EPA-822-R-05-006

US EPA (2006) Reregistration eligibility decision (RED) diazinon. US Environmental Protection Agency, Office of Prevention, Pesticides and Toxic Substances, Office of Pesticide Programs, US Government Printing Office, Washington, DC, EPA 738-R-04-006

US EPA (2007) Risks of diazinon use to the federally listed California red legged frog (*Rana aurora draytonii*). Environmental Fate and Effects Division, Office of Pesticide Programs, Washington, DC

US EPA (2011) Office of Pesticide Programs' Aquatic Life Benchmarks. http://www.epa.gov/oppefed1/ecorisk_ders/aquatic_life_benchmark.htm#benchmarks. Accessed 21 Nov 2011

Wan MT, Szeto SY, Price P (1994) Organophosphorus insecticide residues in farm ditches of the lower Fraser Valley of British Columbia. Journal of Environmental Science and Health Part B Pesticides Food Contaminants and Agricultural Wastes 29(5):917–949

Wofford P, Segawa R, Schreider J (2009) Pesticide air monitoring in Parlier, CA. California Environmental Protection Agency, Department of Pesticide Regulation, Sacramento, CA. http://www.cdpr.ca.gov/docs/envjust/pilot_proj/parlier_final.pdf. Accessed 26 July 2010

Wolfe NL, Mingelgrin U, Miller GC (1990) Abiotic transformations in water, sediments, and soil. In: Cheng HH (ed) Pesticides in soil environment, vol 2. SSSA, Madison, WI, pp 103–168

Zabik JM, Seiber JN (1993) Atmospheric transport of organophosphate pesticides from California's Central Valley to the Sierra Nevada Mountains. J Environ Qual 22:80–90

Zhang X, Starner K (2011) Analysis of diazinon agricultural use in regions of frequent surface water detections. California Department of Pesticide Regulation, Sacramento, CA. Available via CDPR: http://www.cdpr.ca.gov/docs/emon/pubs/ehapreps/analysis_memos/diazinon_anal_zhang_starner_final_091311.pdf. Accessed 21 Nov 2011

Zhou Q, Sun X, Gao R, Hu J (2011) Mechanism and kinetic properties for OH-initiated atmospheric degradation of the organophosphorus pesticide diazinon. Atmos Environ 45:3141–3148

ERRATUM TO

Diazinon–Chemistry and Environmental Fate: A California Perspective

Vaneet Aggarwal, Xin Deng, Atac Tuli, and Kean S. Goh

DOI 10.1007/978-1-4614-5577-6_6

In the paper "Diazinon-Chemistry and Environmental Fate: A California Perspective" by Aggarwal et al. (2013), some of the statements are incorrect. The goal of this erratum is to make corrections on those statements. The corrections are as follows:

1. The last sentence on Page 111 and its continuation on page 113 of Use Profile for Diazinon in California section reads as *"The effect of phasing diazinon out of agricultural use was more pronounced in 2009; in 2009, the total treated area was reduced to 56,905 ha, only 55 and 67% of that treated in 2008 and 2007, respectively."* It should be read as *"However, in 2009, the total agricultural area treated was reduced to 56,905 ha, only 55% and 67% of that treated in 2008 and 2007, respectively."*
2. The sentence at the fourth line of Summary Section on page 134 read as *"In 2004, residential use of diazinon was discontinued, and in 2009, diazinon was phased out of all agricultural uses. Consequently, the amounts of diazinon applied have been drastically decreased."* should be read as *"In 2004, residential use of diazinon was discontinued; as a result, the total amount applied has drastically decreased."*

Reference

Aggarwal V, Deng X, Tuli A, Goh KS (2013) Diazinon-Chemistry and Environmental Fate: A California Perspective. Reviews of Environmental Contamination and Toxicology 223: 107–140

The online version of the original chapter can be found at
http://dx.doi.org/DOI 10.1007/978-1-4614-5577-6_5

INDEX

ACC (1-aminocyclopropane-1-carboxylic acid) deaminase, role in plant stress reduction, **223**: 44
ACC, plant root ethylene effect (diag.), **223**: 45
Acute toxicity to aquatic species, diazinon, **223**: 130
Acute toxicity to freshwater aquatic species, diazinon (table), **223**: 131
Agricultural vs. urban use in California, diazinon (table), **223**: 113
Air behavior, diazinon, **223**: 126
Air contaminants, categories, **223**: 3
Air contaminants, gases, **223**: 5
Air contaminants, health effects, **223**: 5
Air contaminants, heavy metals, **223**: 7
Air contaminants, persistent organic pollutants (POPs), **223**: 7
Air contaminants, Santiago, Chile, **223**: 1ff.
Air contaminants, statistical distributions, **223**: 8
Air contaminants, suspended solid particles, **223**: 7
Air contamination studies, based on Birnbaum-Saunder's models (table), **223**: 13
Air contamination studies, Santiago Chile (table), **223**: 13
Air contamination, Santiago, Chile, **223**: 12
Air contamination, worldwide (table), **223**: 9-11
Air degradation pathway, diazinon (diag.), **223**: 127
Air monitoring locations, Santiago, Chile (illus.), **223**: 15

Air monitoring, diazinon in California, **223**: 128
Air pollutant characteristics, Santiago, Chile, **223**: 17
Air pollutant concerns, the Americas, **223**: 2
Air pollutant effects, Santiago, Chile, **223**: 15
Air pollutant hazards, Chilean guidelines (table), **223**: 16
Air pollutant monitoring, Santiago, Chile, **223**: 14
Air pollutants, maximum permitted levels (table), **223**: 16
Air pollutants, particulate matter, **223**: 3
Air pollution data, statistical analysis, **223**: 17
Air pollution data, statistical models, **223**: 3
Air pollution study, Santiago, Chile, **223**: 16
Air pollution, health effects, **223**: 2
Air pollution, Santiago, Chile, **223**: 2
American kestrel, methyl mercury toxicity, **223**: 64
Antibacterial properties, silver compounds, **223**: 89
Apoptosis, ROS (reactive oxygen species) mode of action for nanosilver, **223**: 87
Aquatic environments, diazinon persistence, **223**: 125
Aquatic life-benchmarks, diazinon, **223**: 133
Aquatic organism toxicity, nanosilver, **223**: 83
Aquatic photolysis, diazinon, **223**: 126
Aquatic species acute toxicity, diazinon (table), **223**: 131
Aquatic species chronic toxicity, diazinon (table), **223**: 132
Aquatic species toxicity, diazinon, **223**: 130

Aquatic species toxicity, nanosilver, **223**: 93
Aquatic toxicity, diazinon, **223**: 129
Aquatic-life benchmark values, diazinon (table), **223**: 133
Atmospheric pollutants, Gaussian statistical modeling, **223**: 8
Atmospheric pollutants, worldwide (table), **223**: 9–11
Autocorrelation analysis, Santiago, Chile, air pollution data, **223**: 18
Avian species in China, compared (table), **223**: 56
Avian species in China, methyl mercury levels (diag.), **223**: 73
Avian species in China, TRV (toxicity reference value) & TRC (tissue residue criteria) derivation, **223**: 66
Avian species toxicity thresholds, methyl mercury, **223**: 65
Avian species toxicity, methyl mercury, **223**: 58
Avian species toxicity, selecting data for protection, **223**: 56
Avian sub-chronic and chronic toxicity, methyl mercury (table), **223**: 59–61
Avian wildlife protection in China, methyl mercury, **223**: 53 ff.
Avian wildlife TRV uncertainty factors, methyl mercury (table), **223**: 66
Avian wildlife TRVs & TRCs, methyl mercury (table), **223**: 67
Avian wildlife, Chinese species to protect, **223**: 55

Bacteria, plant growth effect, **223**: 40
Bacterial cell-wall interaction, silver nanoparticles, **223**: 90
Bioaccumulation, diazinon, **223**: 132
Bioavailability in soil, metals, **223**: 37
Bioavailability of metals, hyperaccumulator plant effect, **223**: 39
Biological effects, nanosilver, **223**: 81 ff.
Bioremediation, definition, **223**: 35
Birnbaum-Saunder's models, air contaminant distribution, **223**: 12

California agricultural counties, diazinon water monitoring data (table), **223**: 124
California annual insecticide use, diazinon (illus.), **223**: 111
California counties, highest diazinon use, **223**: 111
California counties, major diazinon use (table), **223**: 112
California top crop uses, diazinon (diag.), **223**: 113
California urban watersheds, diazinon residues (table), **223**: 124
California use by application method, diazinon (diag.), **223**: 114
California use profile, diazinon, **223**: 107, 110
California, county map (illus.), **223**: 110
Chemistry, diazinon, **223**: 107ff.
Chile, Santiago air contamination, **223**: 1ff.
Chilean guidelines, air pollutant hazards (table), **223**: 16
China, Hg as environmental contaminant, **223**: 54
China, TRVs for avian wildlife, **223**: 53 ff.
Chinese bird species, characterized (table), **223**: 56
Chinese bird species, methyl mercury levels (diag.), **223**: 73
Chinese bird species, most vulnerable to xenobiotics, **223**: 55
Chinese fish residues, methyl mercury (diag.), **223**: 70
Chronic toxicity to aquatic species, diazinon, **223**: 130
Chronic toxicity to freshwater aquatic species, diazinon (table), **223**: 132
CO, air contaminant, **223**: 5
Common loon, methyl mercury toxicity, **223**: 63
Criteria setting for Chinese birds, toxicity data selection, **223**: 56

Daphnids, silver nanoparticle uptake, **223**: 96
Daphnids, silver toxicity, **223**: 93
Death rate in Americas, from air pollution, **223**: 3
Deriving TRC & TRV values, methodology, **223**: 57
Diazinon annual use, California urban vs. total (diag.), **223**: 111
Diazinon application methods, total use in California (diag.), **223**: 114
Diazinon effects, microbial populations, **223**: 121
Diazinon in California, air monitoring, **223**: 128
Diazinon insecticide, uses, **223**: 107
Diazinon persistence, aquatic environments, **223**: 125

Diazinon residues, California urban watersheds (table), **223**: 124
Diazinon residues, water, **223**: 122
Diazinon soil degradation, influencing factors, **223**: 120
Diazinon soil mobility, temperature effect, **223**: 118
Diazinon use volume, major California counties (table), **223**: 112
Diazinon water monitoring data, California agricultural counties (table), **223**: 124
Diazinon, acute & chronic toxicity to aquatic species, **223**: 130
Diazinon, air degradation pathway (diag.), **223**: 127
Diazinon, aquatic life-benchmark and water quality criteria, **223**: 133
Diazinon, aquatic photolysis, **223**: 126
Diazinon, aquatic toxicity, **223**: 129
Diazinon, aquatic-life benchmark values (table), **223**: 133
Diazinon, behavior in air, **223**: 126
Diazinon, bioaccumulation, **223**: 132
Diazinon, California use profile, **223**: 107, 110
Diazinon, chemical structure (illus.), **223**: 108
Diazinon, chemistry & environmental fate, **223**: 107ff.
Diazinon, earthworm toxicity, **223**: 122
Diazinon, environmental fate, **223**: 114
Diazinon, ground- & surface-water residues, **223**: 122-3
Diazinon, groundwater detections, **223**: 108
Diazinon, mode of toxic action, **223**: 129
Diazinon, partition coefficient, **223**: 115
Diazinon, photolysis, **223**: 120
Diazinon, physicochemical properties (table), **223**: 109
Diazinon, physicochemical properties, **223**: 108
Diazinon, soil dissipation, **223**: 119
Diazinon, soil leaching, **223**: 117
Diazinon, soil microbial degradation, **223**: 119
Diazinon, soil mineralization, **223**: 118
Diazinon, soil photolysis pathway (diag.), **223**: 121
Diazinon, soil sorption, **223**: 115
Diazinon, soil-sediment degradation, **223**: 114
Diazinon, top crop uses in California (diag.), **223**: 113
Diazinon, urban vs. agricultural uses in California, **223**: 113
Diazinon, volatilization loss, **223**: 127
Diazinon, water quality criteria (table), **223**: 133

Earthworm toxicity, diazinon, **223**: 122
Environmental fate, diazinon, **223**: 107ff., 114
Enzyme interactions, nanosilver toxicity, **223**: 98
Exceedance probabilities, Santiago, Chile air pollutant data, **223**: 23

Fish residues in China, methyl mercury (diag.), **223**: 70
Fish toxicity mechanism, silver, **223**: 93

Gases, air contaminants, **223**: 5
Gaussian statistical modeling, atmospheric pollutants, **223**: 8
Great egret, methyl mercury toxicity, **223**: 62
Groundwater residues, diazinon, **223**: 122
Guidelines, air pollution in Santiago, Chile, **223**: 15

Health effects, air contaminants, **223**: 5
Health effects, air pollution, **223**: 2
Heavy metal induction, reactive oxygen species, **223**: 40
Heavy metal phytoremediation, rhizobacteria role, **223**: 33ff.
Heavy metal plant stress, PGPR (plant growth-promoting rhizobacteria) counter effect, **223**: 42
Heavy metals, air contaminants, **223**: 7
Heavy metals, exposures & interactions, **223**: 34
Heavy metals, remediation, **223**: 35
Heavy metals, role in plants, **223**: 34
Heavy metals, soil bioavailability, **223**: 37
Heavy metals, toxic mechanisms in plants, **223**: 34
Heavy-metal remediation, PGPR role, **223**: 43
Human health effects, relation to statistical information, **223**: 24
Human health effects, Santiago, Chile air pollution, **223**: 15
Hyperaccumulator plants, in phytoremediaton, **223**: 37
Hyperaccumulator plants, metal bioavailability effect, **223**: 39

In vitro toxicity, silver, **223**: 83, 92
International standards, maximum concentrations for air pollutants (table), **223**: 16
Inversion effects, Santiago, Chile air pollution, **223**: 14

Log-normal distribution, fitting air pollutant data **223**: 12

Mallard, methyl mercury toxicity, **223**: 58
Mercury, as Chinese environmental contaminant, **223**: 54
Mercury, as global environmental contaminant, **223**: 54
Metal pollutants, sources, **223**: 33
Metal-binding properties, metallothionein proteins, **223**: 39
Metallothionein proteins, metal-binding properties, **223**: 39
Metals, plant detoxification mechanisms, **223**: 39
Metals, plant uptake and transport, **223**: 38
Metals, soil bioavailability, **223**: 37
Methyl mercury avian toxicity, species sensitivity distribution (diag.), **223**: 68
Methyl mercury in China, bird protection criteria, **223**: 53 ff.
Methyl mercury levels, avian species in China (diag.), **223**: 73
Methyl mercury toxicity, common Chinese wildfowl species, **223**: 58-64
Methyl mercury, avian species toxicity thresholds (diag.), **223**: 65
Methyl mercury, avian wildlife TRV uncertainty factors (table), **223**: 66
Methyl mercury, avian wildlife TRVs & TRCs (table), **223**: 67
Methyl mercury, bird reproductive effects, **223**: 62
Methyl mercury, bird toxicity, **223**: 58
Methyl mercury, fish concentrations in China (diag.), **223**: 70
Methyl mercury, sub-chronic & chronic avian toxicity (table), **223**: 59-61
Methyl mercury, toxicity thresholds for avian species (table), **223**: 67
Methyl mercury, wildlife contaminant, **223**: 55
Microbial population effects, diazinon, **223**: 121
Mode of action for nanosilver, apoptosis via ROS, **223**: 87
Mode of toxic action, diazinon, **223**: 129
Modes of action possibilities, nanosilver (diag.), **223**: 88
Modes of action, nanosilver, **223**: 81 ff.

Nanoparticle aggregation, toxicity effect, **223**: 94

Nanoparticle silver uptake, daphnids, **223**: 96
Nanoparticle toxicity role, silver ions, **223**: 97
Nanoparticles, description, **223**: 81
Nanosilver environmental behavior, influencing factors, **223**: 82
Nanosilver toxicity in vivo, effective concentration (table), **223**: 95
Nanosilver toxicity, causes, **223**: 98
Nanosilver toxicity, influencing factors, **223**: 82
Nanosilver toxicity, ROS mode of action (table), **223**: 85
Nanosilver toxicity, vs. silver ion toxicity, **223**: 89
Nanosilver, applications, **223**: 82
Nanosilver, aquatic organism toxicity, **223**: 83
Nanosilver, aquatic species toxicity, **223**: 93
Nanosilver, biological effects & modes of action, **223**: 81 ff.
Nanosilver, possible modes of action (diag.), **223**: 88
Nanosilver, size-dependent toxicity, **223**: 84
NO_2, air contaminant, **223**: 6

O$_3$, air contaminant, **223**: 6

Particulate matter, air pollutants, **223**: 3
Partition coefficient, diazinon, **223**: 115
Persistent organic pollutants (POPs), air contaminants, **223**: 7
PGPR (plant growth-promoting rhizobacteria), functions, **223**: 35
PGPR, counter heavy-metal plant stress, **223**: 42
PGPR, heavy metal phytoremediation, **223**: 33ff.
PGPR, in plant heavy-metal remediation, **223**: 43
PGPR, in rhizosphere, **223**: 40
PGPR, plant growth-regulating compounds (table), **223**: 41
Photolysis, diazinon, **223**: 120
Photolytic breakdown, diazinon pathway in soil (diag.), **223**: 121
Physicochemical properties, diazinon (table), **223**: 109
Physicochemical properties, diazinon, **223**: 108
Phytodegradation, phytoremediation type, **223**: 37
Phytoextraction, plant bioremediation, **223**: 35, 36
Phytoextraction, uptake process, **223**: 38

Index

Phytoremediation of heavy metals, rhizobacteria role, **223**: 33ff.
Phytoremediation role, PGPR, **223**: 35
Phytoremediation role, rhizobacteria, **223**: 35
Phytoremediation success, variables, **223**: 35
Phytoremediation, defined, **223**: 36
Phytoremediation, hyperaccumulator plant role, **223**: 37
Phytoremediation, types defined, **223**: 36
Phytostabilization, type of phytoremediation, **223**: 36
Phytostimulation, type of phytoremediation, **223**: 36
Phytovolatilization, type of phytoremediation, **223**: 36
Plant detoxification mechanisms, metals, **223**: 39
Plant growth-promoting rhizobacteria (PGPR), phytoremediation role, **223**: 35
Plant heavy-metal remediation, PGPR synergy, **223**: 43
Plant root ethylene effect, ACC (diag.), **223**: 45
Plant stress reduction, ACC role, **223**: 44
Plant toxicity effects, reactive oxygen species, **223**: 39
Plant toxicity mechanisms, heavy metals, **223**: 34
Plant toxicity, heavy metals, **223**: 34
Plant uptake & transport, metals, **223**: 38
Plants, role of heavy metals, **223**: 34
Plant-soil interactions, heavy metals, **223**: 34
PM10 98th percentile, air pollutant data in Santiago, Chile (table), **223**: 24
PM10 air pollutant data analysis, Santiago, Chile, **223**: 20
PM10 air pollutant data, Santiago, Chile locations (table), **223**: 21
PM10 air pollutant levels, Santiago, Chile monitoring stations (table), **223**: 18
PM10 data analysis, Santiago, Chile, **223**: 17
PM10 level histograms, Santiago, Chile air pollution (diag.), **223**: 19
PM10 levels time series, Santiago, Chile air pollution (diag.), **223**: 19
Probability plots, PM10 air pollutant data for Santiago, Chile (diags.), **223**: 23, 24

Reactive oxygen species (ROS), heavy metal induction, **223**: 40
Regulating air pollutants, Santiago, Chile, **223**: 4

Reproductive effects in birds, methyl mercury, **223**: 62
Residue criteria for methyl mercury, bird protection in China, **223**: 53 ff.
Rhizobacteria, in heavy metal phytoremediation, **223**: 33ff.
Rhizobacteria, role in phytoremediation, **223**: 35
Rhizofiltration, type of phytoremediation, **223**: 37
Rhizovolatilization, type of phytoremediation, type of phytoremediation, **223**: 36
ROS generation, nanosilver toxicity role, **223**: 98
ROS, nanosilver toxicity mechanism, **223**: 82
ROS, nanosilver toxicity mode of action (table), **223**: 85
ROS, plant toxicity effects, **223**: 39
ROS, silver nanoparticle-induced toxicity, **223**: 84

Santiago, Chile air pollutant data, exceedance probabilities & percentiles, **223**: 223
Santiago, Chile air pollutant data, PM10 98th percentile (table), **223**: 24
Santiago, Chile air pollution data, autocorrelation analysis, **223**: 18
Santiago, Chile air pollution, human health effects & guidelines, **223**: 15
Santiago, Chile air pollution, inversion effects, (table), **223**: 14
Santiago, Chile air pollution, PM10 level histograms & boxplots (diag.), **223**: 19
Santiago, Chile air pollution, PM10 levels time series (diag.), **223**: 19
Santiago, Chile air pollution, statistical treatment, **223**: 17
Santiago, Chile locations, PM10 air pollutant data, (table), **223**: 21
Santiago, Chile PM10 levels, boxplots (diag.), (table), **223**: 22
Santiago, Chile, air contamination, **223**: 12
Santiago, Chile, air contamination, **223**: 1ff.
Santiago, Chile, air monitoring locations (illus.), **223**: 15
Santiago, Chile, air pollutant characteristics, **223**: 17
Santiago, Chile, air pollutant problems & regulations, **223**: 4
Santiago, Chile, air pollution data, **223**: 16
Santiago, Chile, air pollution monitoring, **223**: 14

Santiago, Chile, PM10 air pollutant levels (table), **223**: 18
Santiego, Chile, PM10 air pollutant data analysis, **223**: 20
Silver compounds, antibacterial properties, **223**: 89
Silver ions, nanoparticle toxicity cause, **223**: 97, 99
Silver nanoparticle toxicity, influencing factors, **223**: 92
Silver nanoparticle toxicity, ROS, **223**: 84
Silver nanoparticle toxicity, zebrafish, **223**: 94
Silver nanoparticles, bacterial cell-wall interaction, **223**: 90
Silver nanoparticles, daphnid uptake, **223**: 96
Silver toxicity mechanisms, freshwater fish & daphnids, **223**: 93
Silver toxicity, in vitro exposure, **223**: 83
Silver toxicity, in vivo exposure, **223**: 92
Silver, basal cell function disruption, **223**: 83
SO$_2$, air contaminant, **223**: 6
Soil bioavailability, heavy metals, **223**: 37
Soil degradation of diazinon, influencing factors, **223**: 120
Soil dissipation, diazinon, **223**: 119
Soil leaching, diazinon, **223**: 117
Soil microbial degradation, diazinon, **223**: 119
Soil mineralization, diazinon, **223**: 118
Soil photolysis pathway, diazinon (diag.), **223**: 121
Soil sorption, diazinon, **223**: 114, 115
Soil-sediment degradation, diazinon, **223**: 114
Species sensitivity distribution, methyl mercury avian toxicity (diag.), **223**: 68
Statistical distribution uses, air pollution data, **223**: 4
Statistical distributions, relation to air pollutant data, **223**: 3, 8
Statistical distributions, Santiago, Chile air contaminants, **223**: 1ff.
Statistical model validation, for Santiago, Chile air pollutant data, **223**: 20
Statistical models, air pollution data, **223**: 3
Statistics, relation to adverse human effects, **223**: 24
Surface-water residues, diazinon, **223**: 123
Suspended solid particles, air pollutants, **223**: 7

Temperature effect, diazinon soil mobility, **223**: 118
Thermal inversion effects, Santiago, Chile air pollution, **223**: 14

Tissue residue criteria (TRC), for avian wildlife in China, **223**: 53 ff.
Toxicity effect, nanoparticle aggregation, **223**: 94
Toxicity influencing factors, silver nanoparticles, **223**: 92
Toxicity mechanism for nanosilver, ROS, **223**: 82
Toxicity mechanisms, heavy metals in plants, **223**: 34
Toxicity of nanosilver, effective in vivo concentration (table), **223**: 95
Toxicity of nanosilver, via ROS induction (table), **223**: 85
Toxicity of silver, in vitro exposure, **223**: 83
Toxicity of silver, in vivo exposure, **223**: 92
Toxicity reference values (TRV), avian wildlife protection, **223**: 53 ff.
Toxicity thresholds for avian species, methyl mercury (diag.), **223**: 65
Toxicity thresholds for methyl mercury, species sensitivity curve fitting (table), **223**: 67
Toxicity, heavy metals in plants, **223**: 34
TRC (tissue residue criteria), for avian wildlife in China, **223**: 53 ff.
TRC, for Chinese bird protection, **223**: 53 ff.
TRCs for selected avian species in China, methyl mercury (table), **223**: 67
TRV & TRC derivation, avian species in China, **223**: 66
TRV & TRC values for methyl mercury, reasonableness, **223**: 68
TRV (toxicity reference value), for avian wildlife in China, **223**: 53 ff.
TRV and TRC values, derivation methods, **223**: 57
TRV setting, role in protecting wildlife, **223**: 55
TRV, for avian wildlife in China, **223**: 53 ff.
TRVs & TRCs, uncertainty evaluation, **223**: 72
TRVs for selected avian species in China, methyl mercury (table), **223**: 67

Urban vs. agricultural uses in California, diazinon (table), **223**: 113

Volatile organic compounds, air contaminants, **223**: 6
Volatilization loss, diazinon, **223**: 127

Water monitoring data for California, diazinon (table), **223**: 124
Water quality criteria, diazinon (table), **223**: 133
Water quality criteria, diazinon **223**: 133
Water residues, diazinon, **223**: 122
Watersheds in California, diazinon residues (table), **223**: 124
White ibis, methyl mercury toxicity, **223**: 63

Wildlife contaminant, methyl mercury, **223**: 55
Wildlife protection, TRV role, **223**: 55
Wind inversion effects, Santiago, Chile air pollution, **223**: 14

Zebrafish, silver nanoparticle toxicity, **223**: 94
β_1–β_2 charts, PM10 air pollutant data for Santiago, Chile (diag.), **223**: 22

Printed by Publishers' Graphics LLC
DBT130512.15.12.1